MEDIA AND COMMUNICATIONS - TECHNOLOGIES, POLICIES AND CHALLENGES

DIGITAL BOOKS: COMPETITION AND COMMERCE

MEDIA AND COMMUNICATIONS – TECHNOLOGIES, POLICIES AND CHALLENGES

Additional books in this series can be found on Nova's website at:

https://www.novapublishers.com/catalog/index.php?cPath=23_29&seriesp=Media+and+Communications+-+Technologies%2C+Policies+and+Challenges

Additional E-books in this series can be found on Nova's website at:

https://www.novapublishers.com/catalog/index.php?cPath=23_29&seriespe=Media+and+Communications+-+Technologies%2C+Policies+and+Challenges

MEDIA AND COMMUNICATIONS - TECHNOLOGIES, POLICIES AND CHALLENGES

DIGITAL BOOKS: COMPETITION AND COMMERCE

OLIVER A. HAGEN
EDITOR

Nova
Nova Science Publishers, Inc.

New York

For permission to use material from this book please contact us:
Telephone 631-231-7269; Fax 631-231-8175
Web Site: http://www.novapublishers.com

LIBRARY OF CONGRESS CATALOGING-IN-PUBLICATION DATA

Digital books : competition and commerce / editor, Oliver A. Hagen.
p. cm.
Includes index.
Includes reprinted legislative documents.
ISBN 978-1-61728-299-7 (hardcover)
1. Electronic publishing--Law and legislation--United States. 2. Book industries and trade--Law and legislation--United States. 3. Library materials--Digitization--United States. 4. Fair use (Copyright)--United States. I. Hagen, Oliver A.
KF2750A25 2009
346.7304'82--dc22
2010015593

Published by Nova Science Publishers, Inc. † New York

CONTENTS

PREFACE

Authors and publishers sued Google Inc. in 2005, shortly after Google announced plans to digitize books in the collections of several major libraries, index them in its search engine and allow searchers to view "snippets" of the digitized books. Google's proposed reproduction and display of copyrighted books was not authorized by the rights holders, who alleged that the Google Library Project infringed their copyrights. Google's counterargument, that allowing rights holders to "opt out" of having their books digitized or indexed kept its proposed uses from being infringed upon, or that, if found to be infringing, its proposed uses were fair, raised important questions about reproduction and fair use under copyright law. This book provides background on the Library Project, legal issues raised by digitization and indexing projects, testimony for and against the project, and details of the proposed settlement.

Chapter 1 - The Google Book Search Library Project, announced in December 2004, raised important questions about infringing reproduction and fair use under copyright law. Google planned to digitize, index, and display "snippets" of print books in the collections of five major libraries without the permission of the books' copyright holders, if any. Authors and publishers owning copyrights to these books sued Google in September and October 2005, seeking to enjoin and recover damages for Google's alleged infringement of their exclusive rights to reproduce and publicly display their works. Google and proponents of its Library Project disputed these allegations. They essentially contended that Google's proposed uses were not infringing because Google allowed rights holders to "opt out" of having their books digitized or indexed. They also argued that, even if Google's proposed uses were infringing, they constituted fair uses under copyright law.

The arguments of the parties and their supporters highlighted several questions of first impression. First, does an entity conducting an unauthorized digitization and indexing project avoid committing copyright infringement by offering rights holders the opportunity to "opt out," or request removal or exclusion of their content? Is requiring rights holders to take steps to stop allegedly infringing digitization and indexing like requiring rights holders to use meta-tags to keep search engines from indexing online content? Or do rights holders employ sufficient measures to keep their books from being digitized and indexed online by publishing in print? Second, can unauthorized digitization, indexing, and display of "snippets" of print works constitute a fair use? Assuming unauthorized indexing and display of "snippets" are fair uses, can digitization claim to be a fair use on the grounds that apparently *prima facie* infringing activities that facilitate legitimate uses are fair uses?

These and related questions could remain unanswered, at least for now, because Google and the rights holders seek to settle the litigation over the Google Book Search Library Project. A settlement agreement was originally proposed on October 28, 2008. However, although the court granted preliminary approval of this agreement on November 17, 2008, opponents of the settlement raised concerns about Google's potential monopolization of book searching, the treatment of orphan works, protection of searchers' privacy, the rights of foreign authors, and related issues in amicus briefs and other court filings. Responding to such concerns, the parties proposed a revised settlement agreement on November 13, 2009. The court granted preliminary approval of the revised agreement on November 19, 2009, and scheduled a final fairness hearing on it for February 18, 2010. Class members have until January 28, 2010, to opt into, opt out of, or object to the revised agreement. The court will consider any objections, as well as conduct an independent review of the proposed agreement, in determining whether to grant final approval.

Chapter 2 - Orphan works are copyrighted works whose owners are difficult or impossible to identify and/or locate. Orphan works are perceived to be inaccessible because of the risk of infringement liability that a user might incur if and when a copyright owner subsequently appears. Consequently, many works that are, in fact, abandoned by owners are withheld from public view and circulation because of uncertainty about the owner and the risk of liability.

In 2006, at the request of Congress, the U.S. Copyright Office issued its *Report on Orphan Works* ("Report"). The goal of the Report was to elicit public comment and evaluate the extent of real or perceived problems that content users encounter in their efforts to use these works. The Report defines the problems it identified, and concludes that the problem is indeed real and should be addressed legislatively. It analyzes stakeholders' views on the issue and constraints on solutions imposed by the structure of U.S. copyright law and international copyright obligations. The Report sets forth a proposal to amend the Copyright Act by adding a provision that would limit liability for infringing use of orphan works when, prior to use, a user performs a reasonably diligent search for the copyright owner and provides attribution to the author and copyright owner, if possible. In some instances, when copyright infringement is made without commercial advantage and the user ceases infringement promptly after receiving notice thereof, no monetary relief would be available.

Adopting many of the suggestions of the Copyright Office, the Orphan Works Act of 2006 was introduced in the 109th Congress, second session (H.R. 5439). This bill was later incorporated into an omnibus copyright bill, appearing as Title II of The Copyright Modernization Act of 2006 (H.R. 6052). However, the bill was not addressed by the end of that Congress's adjournment. The bill would have implemented a limitation on monetary damage liability for specified infringement of orphan works, but took a more detailed approach than the Report's original proposals in establishing requirements for such liability limitations, such as articulating standards for a "reasonably diligent search." The bill would also have directed the Copyright Office to study and report on the implementation of the new orphan works amendment, and to study and make recommendations for a "small claims" procedure to address copyright infringement.

Legislation addressing the orphan works issue was reintroduced in the 110th Congress: the Orphan Works Act of 2008 (H.R. 5889) and the Shawn Bentley Orphan Works Act of 2008 (S. 2913). The two bills resembled the Orphan Works Act of 2006, although there were substantial differences from that earlier legislation and even between themselves. These additional or revised provisions were added in part to address concerns raised by photographers, illustrators, and other visual artists, as well as textile and home furnishing manufacturers. While S. 2913 passed the Senate, H.R. 5889 did not make it out of the House Judiciary Committee.

This chapter surveys the findings and conclusions in the U.S. Copyright Office's *Report on Orphan Works* and analyzes the orphan works bills that were considered by the 109th and 110th Congresses. No legislation relating to orphan works has yet been introduced in the 111th Congress as of the date of this chapter. However, the outcome of the Google Book Search class action lawsuit (and its pending settlement) may potentially affect future orphan works legislation.

Chapter 3 - This chapter is edited and excerpted testimony by Paul Aiken before the House Judiciary Committee on Banking on September 10, 2009.

Chapter 4 - This chapter is edited and excerpted testimony by Marybeth Peters before the House Judiciary Committee on Banking on September 10, 2009.

Chapter 5 - This chapter is edited and excerpted testimony by David A. Balto before the House Judiciary Committee on Banking on September 10, 2009.

Chapter 6 - This chapter is edited and excerpted testimony by David Drummond before the House Judiciary Committee on Banking on September 10, 2009.

Chapter 7 - This chapter is edited and excerpted testimony by Paul Misener before the House Judiciary Committee on Banking on September 10, 2009.

Chapter 8 - This chapter is edited and excerpted testimony by Randal C. Picker before the House Judiciary Committee on Banking on September 10, 2009.

Chapter 9 - This chapter is edited and excerpted testimony by Marc Maurer before the House Judiciary Committee on Banking on September 10, 2009.

Chapter 10 - This chapter is edited and excerpted testimony by John M. Simpson before the House Judiciary Committee on Banking on September 10, 2009.

In: Digital Books: Competition and Commerce
Editor: Oliver A. Hagen ISBN: 978-1-61728-299-7

Chapter 1

THE GOOGLE LIBRARY PROJECT: IS DIGITIZATION FOR PURPOSES OF ONLINE INDEXING FAIR USE UNDER COPYRIGHT LAW?

Kate M. Manuel

SUMMARY

The Google Book Search Library Project, announced in December 2004, raised important questions about infringing reproduction and fair use under copyright law. Google planned to digitize, index, and display "snippets" of print books in the collections of five major libraries without the permission of the books' copyright holders, if any. Authors and publishers owning copyrights to these books sued Google in September and October 2005, seeking to enjoin and recover damages for Google's alleged infringement of their exclusive rights to reproduce and publicly display their works. Google and proponents of its Library Project disputed these allegations. They essentially contended that Google's proposed uses were not infringing because Google allowed rights holders to "opt out" of having their books digitized or indexed. They also argued that, even if Google's proposed uses were infringing, they constituted fair uses under copyright law.

The arguments of the parties and their supporters highlighted several questions of first impression. First, does an entity conducting an unauthorized digitization and indexing project avoid committing copyright infringement by offering rights holders the opportunity to "opt out," or request removal or exclusion of their content? Is requiring rights holders to take steps to stop allegedly infringing digitization and indexing like requiring rights holders to use meta-tags to keep search engines from indexing online content? Or do rights holders employ sufficient measures to keep their books from being digitized and indexed online by publishing in print? Second, can unauthorized digitization, indexing, and display of "snippets" of print works constitute a fair use? Assuming unauthorized indexing and display of "snippets" are fair uses, can digitization claim to be a fair use on the grounds that apparently *prima facie* infringing activities that facilitate legitimate uses are fair uses?

These and related questions could remain unanswered, at least for now, because Google and the rights holders seek to settle the litigation over the Google Book Search Library Project. A settlement agreement was originally proposed on October 28, 2008. However, although the court granted preliminary approval of this agreement on November 17, 2008, opponents of the settlement raised concerns about Google's potential monopolization of book searching, the treatment of orphan works, protection of searchers' privacy, the rights of foreign authors, and related issues in amicus briefs and other court filings. Responding to such concerns, the parties proposed a revised settlement agreement on November 13, 2009. The court granted preliminary approval of the revised agreement on November 19, 2009, and scheduled a final fairness hearing on it for February 18, 2010. Class members have until January 28, 2010, to opt into, opt out of, or object to the revised agreement. The court will consider any objections, as well as conduct an independent review of the proposed agreement, in determining whether to grant final approval.

INTRODUCTION

Authors and publishers sued Google Inc. in 2005, shortly after Google announced plans to digitize books in the collections of several major libraries, index them in its search engine (http://www.google), and allow searchers to view "snippets" of the digitized books. Google's proposed reproduction and display of copyrighted books was not authorized by the rights holders, who alleged that the Google Library Project infringed their copyrights. Google's counterarguments—that allowing rights holders to "opt out" of having their books digitized or indexed kept its proposed uses from being infringing, or that, if found to be infringing, its proposed uses were fair—raised important questions about reproduction and fair use under copyright law. Namely, does an entity engaged in unauthorized digitization and indexing avoid committing copyright infringement by offering rights holders the opportunity to request removal or exclusion of their content? And, assuming unauthorized indexing and display of "snippets" are fair uses, can digitization claim to be a fair use on the grounds that apparently *prima facie* infringing activities that facilitate legitimate uses are fair uses? The revised proposed settlement agreement between Google and rights holders could mean that litigation over the Library Project will not help to answer these questions. However, final court approval of the revised agreement is pending, and future digitization and indexing projects may raise similar questions.

This chapter provides background on the Library Project, legal issues raised by digitization and indexing projects, and the proposed settlement. It will be updated as developments warrant. It supersedes CRS Report RS22356, *The Google Book Search Project: Is Online Indexing a Fair Use Under Copyright Law?,* by Robin Jeweler.

THE GOOGLE LIBRARY PROJECT

In December 2004, Google initiated its Library Project by announcing partnerships with five libraries.[1] Under the partnership agreements, the libraries would allow Google to digitize the print books in their collections, and Google would (1) index the contents of the books; (2)

display at least "snippets" of the books among its search results; and (3) provide partner libraries with digital copies of the print books in their collections.[2] Google and its partners never planned to make the full text of any digitized and indexed books that are still within their terms of copyright protection available to searchers.[3] Rather, by digitizing and indexing books, Google and its partners sought to make the contents of print books more accessible to searchers, who could potentially buy or borrow books after seeing "snippets" of them among the results of Google searches.[4] Google also intended to sell advertising "keyed" to results lists incorporating the digitized books.[5]

Google's Library Project was itself part of a larger initiative initially known as Google Print and later renamed Google Book Search.[6] The Google Partner Program was also part of this initiative.[7] The Partner Program allowed authors and publishers to submit copies of their books for indexing in Google's search engine.[8] However, because rights holders affirmatively chose to have their books digitized or indexed through the Partner Program, the Program was not subject to allegations of copyright infringement like those made against the Library Project.

THE LITIGATION AND THE PARTIES' POSITIONS

Authors and publishers objected to the Google Library Project from its inception on the grounds that it infringed their copyrights.[9] Generally, copyrights in books initially vest in the books' authors.[10] Many authors later transfer their copyrights to publishers under contract in exchange for payment and the publisher's manufacturing and selling copies of the book.[11] Regardless of whether they are the books' authors or publishers, however, copyright holders have exclusive rights "to reproduce the copyrighted work in copies," or, in the case of literary works such as books, "to display the copyrighted work publicly."[12] The authors and publishers who objected to the Library Project claimed that Google infringed these exclusive rights by making digital copies of print books and presenting snippets from the digitized books without rights holders' permission. [13] Google initially responded to these concerns by allowing rights holders who did not want their books included in Google Book Search to "opt out."[14] If rights holders notified Google, Google would ensure that digitized versions of their books were not included in its database.

The ability to "opt out" of the Library Project did not satisfy authors and publishers, however. They sued to enjoin Google's digitization and indexing and to recover monetary damages for Google's alleged copyright infringement. In September 2005, the Authors Guild filed a class action suit in U.S. District Court for the Southern District of New York on behalf of "all persons or entities that hold the copyright to a literary work that is contained in the library of the University of Michigan."[15] Shortly thereafter, five publishing companies also sued in the Southern District of New York.[16] The suits were consolidated, and additional plaintiffs, including the Association of American Publishers, joined the suit. Because the consolidated case was a class action, the court must approve any settlement of it.[17]

In responding to the suit, Google essentially contended that its conduct was not infringing because it gave rights holders the opportunity to "opt out" of having their books digitized and indexed.[18] Google also claimed that, even if a court found its conduct to be infringing, this conduct represented a fair use of the rights holders' works.[19] Google and supporters of its

Library Project specifically cited the decision by the U.S. Court of Appeals for the Ninth Circuit in *Kelly v. Arriba Soft Corporation* as support for the proposition that the indexing activities of Internet search engines constitute fair uses.[20]

LEGAL ISSUES RAISED BY THE LITIGATION

The litigation over the Google Library Project raised important questions about infringing reproduction and fair use under copyright law. Namely, can an entity engaging in unauthorized digitization and indexing avoid liability for copyright infringement by offering rights holders the opportunity to request removal or exclusion of their content from its database? And, assuming unauthorized indexing and display of "snippets" of digitized works are fair uses, can digitization itself claim to be a fair use on the grounds that apparently *prima facie* infringing activities that facilitate legitimate uses are fair uses? These questions will arguably persist, and their answers remain important, even if the parties ultimately settle the litigation over the Library Project.

"Opt Out" Programs and Liability for Infringement

Google's first line of defense against the authors and publishers was essentially that it was not liable for copyright infringement because it gave rights holders the opportunity to "opt out" of having their works digitized and indexed. In making this argument, Google relied on the related claim that no one would conduct multi-library digitization and indexing projects like the Library Project if they had to clear the copyrights for every book with the rights holders.[21] Identifying and locating the rights holder(s) for one book can be difficult enough, supporters of the Google Library Project noted, without repeating this process millions of times, as would be necessary with a major library collection.[22] The publishers, in contrast, noted that Google's offer to let rights holders "opt out" of having their books digitized and indexed "stands copyright law on its head."[23] They argued that one cannot generally announce one's intention to infringe multiple copyrighted works and collectively offer rights holders the opportunity *not* to have their works infringed.[24]

It is impossible to predict what a court would find based on such arguments, and this chapter does not attempt to do so. This chapter does, however, highlight some of the considerations that could factor in the court's consideration of the issue. On the one hand, the requirement that a copyright owner act affirmatively to stop non-willful infringement is not without precedent. The "notice and takedown" procedures of the Digital Millennium Copyright Act (DMCA),[25] for example, require content owners to notify Internet Service Providers (ISPs) of the existence of infringing content and can immunize ISPs from liability for infringement when they serve as "passive conduits" for infringing content transmitted by third parties.[26] Similarly, at least one court has found that content owners are responsible for taking affirmative measures, such as using meta-tags within the computer code of a Web page, to prevent Internet search engines from automatically indexing and displaying their content.[27] On the other hand, plaintiffs could argue that comprehensive digitization projects, like that proposed by Google, willfully infringe copyright[28] and differ from the "passive conduits" protected by the DMCA. Likewise, rights holders in print books could argue that their situations differ from that of Web page authors because Google had to digitize their books before indexing them. They could claim that they took sufficient affirmative measures to protect their works by not making them available for free on the Web.[29]

Digitization, Indexing, and Display as Fair Uses

Google also attempted to defend against the rights holders' allegations of copyright infringement by claiming that the Library Project, if found to be infringing, constituted a fair use.[30] The "fair use" exemption within copyright law limits rights holders' exclusive rights by providing that uses for "certain purposes"—including, but not limited to, criticism, comment, news reporting, teaching, scholarship, and research—do not infringe copyright *even if* they are made without the rights holders' consent.[31] In determining whether challenged conduct constitutes a fair use, a court considers the following factors, which were developed under the common law and later codified in the Copyright Act of 1976:

1. the purpose and character of the use, including whether such use is of a commercial nature or is for nonprofit educational purposes;
2. the nature of the copyrighted work;
3. the amount and substantiality of the portion used in relation to the copyrighted work as a whole; and
4. the effect of the use upon the potential market for or value of the copyrighted work.[32]

These four factors must not be "treated in isolation, one from another."[33] Rather, "[a]ll are to be explored, and the results weighed together, in light of the purposes of copyright,"[34] which is to "Promote the Progress of Science and useful Arts" and serve the public welfare.[35] Also, because fair use is an "equitable rule of reason" to be applied in light of copyright law's overall purposes, other relevant factors may be considered.[36] The court hearing the case makes findings of fact and assigns relative value and weight to each of the fair use factors. The court can also look to prior cases for guidance even though determining whether a challenged activity constitutes a fair use "calls for a case-by-case analysis."[37]

Although it is impossible to predict what a court would find when confronted with an actual case, and this chapter will not attempt to do so, it does highlight some of the many questions that the Google Library Project raised regarding each of the four statutory "fair use" factors. The report does so in order to illustrate the potential importance of the Library Project—or similar digitization and indexing projects—in establishing the scope of infringing reproduction and fair use under copyright law.

The Purpose and Character of the Use

First, as regards the purpose and character of the use, copyright law generally presumes that commercial uses are not fair,[38] and that transporting a work to a new medium is not a fair use.[39] These presumptions would seem to work against digitization and indexing projects like the Library Project. The Project was implemented by a for-profit corporation that proposed, among other things, to sell ads "keyed" to the digitized content. The Project was also intended to migrate content from print to digital format. These presumptions can, however, be overridden when the use is sufficiently transformative.[40] A copy's use of the original is transformative when the copy does not "merely supersede[]" the original but rather "adds something new, with a further purpose or a different character" to the original.[41]

The transformative nature of the Library Project would arguably be more easily established if it merely indexed books and displayed "snippets" of them. Were Google's uses so limited, it could probably rely on the precedent of two cases from the U.S. Court of Appeals for the Ninth Circuit which found that indexing and abridged displays of copyrighted content were fair uses. In the first case, *Kelly v. Arriba Soft Corporation*, the court held that a company operating a search engine, which had indexed a rights holder's online photographs and displayed "thumbnail" versions of them, was not liable for copyright infringement because its uses were fair.[42] Key to this holding was the court's finding that indexing represented a transformative use of the original photographs. While the original photographs were intended "to inform and to engage the viewer in an aesthetic experience," Arriba used its copies of them for a different function: "improving access to information on the internet."[43] The court also emphasized that Arriba indexed and displayed "thumbnail" versions of the photographs.[44] The thumbnails had much lower resolution than the originals and thus could not substitute for them because "enlarging them sacrifices their clarity."[45] The Ninth Circuit reached a similar conclusion in *Perfect 10, Inc. v. Amazon.com, Inc.*[46] There, the court also considered a use's benefit to society in finding the use to be transformative. The court noted that "a search engine provides social benefit by incorporating an original work into a new work, namely, an electronic reference tool."[47]

The digitization involved in the Library Project complicates the analysis, however. Admittedly, the prior cases that found indexing and abridged displays of copyrighted content to be fair uses also involved copying of originals.[48] However, in these cases, the copying was of originals posted on the Internet and resulted in copies that were "inferior" to the originals for all purposes except their use in indexing. The first difference is potentially significant because courts have held that rights holders confer limited licenses to copy their content for purposes of indexing and abridged display by posting it on the Internet without taking affirmative measures to prevent copying.[49] The second difference could also be significant because digitized books are arguably superior to print ones when it comes to locating specific information within them.[50]

Because digitization was so central to the Library Project, and arguably could not be directly paralleled to the copying in cases involving indexing and display of Internet materials, Google might have had to rely on the proposition that apparently *prima facie* infringing activities (such as digitization) that facilitate legitimate uses (such as indexing and limited displays) are fair uses. The Supreme Court's decision in *Sony Corporation of America v. Universal City Studios* could arguably provide broad support for this principle.[51] In *Sony*, the Court held that the sale of the video recording machine, which was used to "time shift" broadcast television for personal home viewing, was not contributory copyright infringement.[52] Although the factual underpinnings and legal precedent of *Sony* are not particularly relevant to or controlling in a case like Google's, the *Sony* decision itself stands as a landmark in copyright law demonstrating the willingness of the Court to balance new technological capabilities against traditional principles of copyright law and to recognize new categories of fair use. Many copyright experts saw analogies to the technological considerations inherent in *Sony* in Google's case.[53] Such experts noted that Google's allegedly infringing activity in digitizing print books was incidental to the valid and socially useful function of indexing.

The analogy to *Sony* might not be enough to persuade a court that digitizing for purposes of non- infringing indexing constitutes a fair use, however. Digitizing and indexing print books are arguably far removed from making and selling devices that consumers use to record broadcast television programming and replay it later. Additionally, courts have shown little inclination to recognize categories of judicially created fair uses other than time shifting. In *UMG Recordings v. MP3 .com, Inc.*, for example, a U.S. district court rejected out-of-hand the defendant's proffered fair use defense as a justification for unauthorized copying of plaintiffs' audio CDs.[54] The defendant had claimed that its unauthorized copying enabled CD owners to "space shift" because they could access the music on their CDs from any location through MP3.com's subscription service.[55]

The Nature of the Copyrighted Work

Comprehensive digitization and indexing projects, such as the Google Library Project, raise similar questions when the second fair use factor is considered. Projects that digitize library collections potentially encompass diverse types of materials. Some of these materials may be works of fiction, which are among the creative works accorded the highest level of copyright protection.[56] Other materials may be reference books or compendiums of facts, which are afforded the "thinnest" copyright protection.[57] Yet other materials may be nonfiction and mix unprotected ideas with protected expressions of these ideas.[58] This diversity of materials makes possible the arguments of both proponents and opponents of the view that projects like Google Book Search constitute fair uses. The nature of the work can, however, be less important than the purpose and character of the use, at least in situations where the use can be clearly recognized as transformative.[59]

The Amount and Substantiality of the Portion Used

The amount and substantiality of the portion used in relation to the copyrighted work as a whole is another factor that could potentially cut either way in cases involving digitization and indexing projects. As a general rule, "[w]hile wholesale copying does not preclude fair use *per se*, copying an entire work militates against a finding of fair use."[60] Copying entire works can, however, be found to constitute a fair use when doing so is reasonable given the purpose and character of the use.[61] Digitization projects, such as the Google Library Project, would clearly be engaged in wholesale copying, including copying any segments comprising the "heart" of the copied work. [62] The question would thus become whether such wholesale copying was reasonable for an indexing project. Proponents of the project could argue that courts have found copying entire works in order to digitize them reasonable,[63] and that searchers would see only "snippets" of the work in any case. Opponents, in contrast, could argue that, in all cases where courts protected wholesale copying for purposes of indexing, the authors had placed their works online, thereby creating implied licenses for others to copy and index them.[64] Moreover, in at least some of these cases, the copies were deleted after the indexing was completed.[65] In no case did the copier propose to give copies to third parties, as Google did when contracting to provide digital copies of the books in their collections to libraries.

The Effect of the Use Upon the Potential Market or Value of the Work

Finally, digitization and indexing projects could be seen as either promoting or inhibiting the potential markets or values of the copyrighted works. Proponents of digitization could argue that indexing and display of "snippets" of print books increases the markets for the originals by alerting researchers to books on their topics. If researchers purchase books of which they would otherwise have been unaware, the markets for these books could potentially be improved by the unauthorized digitization. Opponents, in contrast, could argue that unauthorized digitization and indexing usurps markets that the rights holders are developing;[66] that viewing "snippets" of print books sometimes can substitute for purchases of them; and that rights holders should be free to determine whether, when, and how their print works are digitized.[67] The outcome of any findings by the court on this factor may hinge upon the degree of harm to their markets that plaintiffs must show. Some courts have required plaintiffs to show only that the markets in which they alleged harm are "likely to be developed,"[68] while others have required proof of actual losses in established markets.[69] The fact that a use is transformative can, however, outweigh even inhibition of or harm to plaintiffs' markets.[70]

ATTEMPTS TO SETTLE THE LITIGATION

Google and the rights holders have sought to settle the litigation over the Google Book Search Library Project since October 2008. They announced a settlement agreement on October 28, 2008. However, this agreement prompted concerns about Google's potential monopolization of book searching, the treatment of orphan works, protection of searchers' privacy, and the rights of foreign authors, among other things. Responding to such concerns, the parties proposed a revised settlement agreement on November 13, 2009. Final approval of the revised agreement is pending. Because the suit is a class action, the judge must ensure that any settlement is "fair, reasonable, and adequate" for the class members before granting final approval.[71]

The Proposed Settlement Agreement of October 2008

On October 28, 2008, Google and the rights holders announced a proposed settlement agreement.[72] Under this agreement, Google would have compensated rights holders for prior and future uses of their work.[73] Google would also have funded the establishment and initial operations of a not-for-profit entity, called the Registry, which would have represented rights holders in negotiating future uses of their content with Google.[74] Google, in turn, would have received a non-exclusive license[75] to (1) "Digitize all Books and Inserts" published before January 5, 2009, and (2) make certain uses of the digitized materials, including displaying "snippets" of them among its search results, subject to the terms of the agreement.[76] By allowing Google to digitize and display books, the agreement would have paved the way for Google to expand Google Book Search, selling subscriptions to institutions and electronic versions of books to individuals.[77] The agreement would also have created certain rights and responsibilities for libraries that allow Google to digitize their books,[78] as well as made certain provisions for subscribers and users of commercialized versions of Google's Book Search database.[79]

The agreement was not to take effect until certain conditions were met, one of which was final court approval of the settlement agreement.[80] Although the court granted preliminary approval of the agreement on November 17, 2008,[81] the final fairness hearing on the settlement was postponed from June 11, 2009, to October 7, 2009,[82] and the parties eventually received permission to submit an amended version of the settlement agreement.[83]

Opposition to the Proposed October 2008 Agreement

Commentators voiced numerous concerns about the proposed October 2008 agreement, some of which recurred in class members' objections and amicus briefs filed with the court. One major concern was that the settlement would have effectively granted Google an exclusive license to digitize the books covered by the agreement and display them to individual users and libraries.[84] The settlement stated that Google's digitization and display rights under the agreement were nonexclusive,[85] and nothing, under the settlement or otherwise, would have precluded another entity from undertaking a digitization and indexing project like Google Book Search.[86] However, some commentators worried that these factors might not prevent Google's effectively monopolizing the book-search field given that the settlement agreement (1) included a clause granting Google the right to any more favorable terms that the Registry negotiated with third-parties over the next 10 years[87] and (2) protected Google from the litigation risks likely to confront those initiating new book-digitization projects. Commentators feared that these factors could significantly diminish the willingness of potential competitors to enter the field and result in a monopolized market where consumers and libraries, in particular, are vulnerable to price increases.[88] On July 2, 2009, the U.S. Department of Justice confirmed that it was conducting an antitrust investigation into the Google settlement because of such concerns.[89] A coalition of potential Google competitors, including Microsoft, Amazon, and Yahoo, also formed to lobby against and otherwise oppose the settlement, in part because a "single organization should not control public access to the books."[90]

A related concern focused upon the settlement's potential effects on "orphan works," or copyrighted works whose owners are difficult or impossible to identify or locate.[91] Some commentators worried that the settlement would grant Google a "unique lock"[92] on orphan works by making the Registry, which was to be created under the agreement, the effective trustee of rights in orphan books.[93] The Registry had already opted to deal with Google regarding these books, and some commentators were concerned that the absence of known or identifiable rights holders effectively meant that these books could not be further used without the agreement of the Registry and/or Google.[94] Commentators concerned about orphan works also suggested that the settlement agreement's treatment of them constituted "a kind of legislation, stepping on congressional prerogatives."[95] The 110th Congress considered, but did not enact, legislation that would have authorized use of orphan works under certain conditions.[96] Similar legislation may be introduced in the 111th Congress. Some commentators felt that any "changes" involving orphan works of the magnitude allegedly made in the settlement agreement ought to be made by Congress, not the federal courts.[97]

Yet other concerns centered upon the settlement agreement's effects on public welfare, especially its provisions regarding public access and user privacy. Several library groups stated that, while they did not oppose the settlement, they were concerned that some provisions could "compromise fundamental library values such as equity of access to information, patron privacy, and intellectual freedom."[98] These groups wanted the court to oversee the pricing of individual subscriptions and the selection of board members for the Registry to ensure that pricing is equitable and the Registry acts in the public interest. Other library groups expressed similar concerns regarding the restrictions on printing and networking imposed on the free subscriptions to Book Search that would be provided to public libraries under the settlement agreement.[99] Relatedly, privacy groups noted that the proposed agreement would allow or require Google to collect extensive information about users, but made few provisions for the protection of this information. As one group stated,

> ... [T]he settlement gives Google the potential, and in some instances the need, to collect substantial quantities of sensitive reader information. Google will also need to share some usage data with the Registry. Specifically, Google will share sales and subscription usage data for calculating and distributing payment to rights holders, market research data concerning various Preview options, and data pertaining to audits and security breaches. Some collection and sharing is of course necessary to effectuate the settlement, but the settlement does not—and should not—require sharing anything other than aggregate data. However, the settlement does not contain a broad restriction on the sharing of user data. The agreement does state that Google cannot be forced to disclose "confidential or personally identifiable information except as compelled by law or valid legal process" in the case of a security breach, but it does not address voluntary disclosure by Google. More generally, it also does not address Google's collection, use, retention, and sharing of user data outside the specific context of a security breach. More formal privacy safeguards would ensure that readers maintain the privacy they have traditionally enjoyed, preserving the right to read anonymously and allowing readers to feel free to access and read books of any sensitive sort.[100]

A challenge to the settlement class itself was also reportedly filed by an attorney and author who alleged that the settlement represented "an abuse of the class-action process" because the settlement class was broad and the named plaintiffs did not adequately represent all class members.[101] This challenge also alleged that the proposed settlement did not adequately compensate authors,[102] a concern which was also raised by writers groups.[103] One writers group noted that "of the $125 million offered by Google, only $45 million [was] for writers," an amount allegedly "way short of the amount needed to compensate authors of millions of books."[104] This group also claimed that the settlement was "grossly unfair" to authors because it required them to "opt out" of Google's proposed uses of their works and interfered in their relationships with publishers.[105]

Support for the Proposed October 2008 Agreement

Supporters of the proposed October 2008 agreement disputed many of the previous claims, as well as noted potential benefits of the project. Some supporters claimed that, far from being anticompetitive, the settlement was pro-competitive because it "would make it easier for competitors to enter the book search market by resolving uncertainty over intellectual property rights, allowing competitors to follow in Google's footsteps."[106] Such claims may get some support from a reported partnership between the University of Michigan and Amazon's subsidiary BookSurge, under which 400,000 titles owned by the university will be made available through print-on-demand.[107] Although the university created the digital versions of some titles itself, others were produced by Google as part of its Library

Project and will be offered through an alleged Google competitor. Other supporters similarly noted that many alleged orphan works are not, in fact, orphans and the settlement will prompt their owners to identify themselves, or reveal that the works are in the public domain.[108] Relatedly, they noted that the agreement would not create an "orphan works monopoly," as some critics feared, because the "scanned books naturally remain in libraries throughout the country, and competitors can use them according to copyright law."[109]

Other supporters noted that the agreement may help "solve disparities in resources available to minorities, the disabled, and the impoverished" by "remov[ing] some of the physical obstacles we have had to acquiring knowledge."[110]Advocates for the persons with visual disabilities asserted that the agreement would "allow blind people to have access to millions of books for the first time in history,"[111] while various civil rights leaders claimed it could democratize knowledge, transform education, and resolve differences in access to information between minority and other communities.[112] The European Union's Telecoms and Information Society Commissioner, Viviane Reding, also reportedly voiced support for the agreement.[113] Reding reportedly asked EU member states to "back similar arrangements in the European Union"; called for an overhaul of EU copyright laws to help expand digitization of orphan works; and stated that "[i]t is good to see that new business models are evolving that could allow bringing more content to an increasing number of consumers."[114]

The Revised Settlement Agreement of November 2009

In response to these and other concerns, the parties proposed a revised settlement agreement on November 13, 2009.[115] While many provisions of the revised agreement are identical to those in the October 2008 agreement, the revised agreement would make several key changes. Among other things, it would

- remove many foreign rights holders from the settlement class by redefining "book" to include only non-U.S. works registered with the U.S. Copyright Office or published in Canada, Australia, or the United Kingdom;[116]
- remove the provision granting Google the right to any more favorable terms that the Registry negotiates with third-parties over the next 10 years;[117]
- explicitly recognize rights holders' right to authorize, through the Registry or otherwise, any third party to use their copyrighted content "in any way, including ways provided for under this Amended Settlement Agreement";[118]
- appoint an independent fiduciary to represent rights holders who have not claimed their works and authorize the fiduciary to spend part of the revenue derived from unclaimed works in searching for their rights holders;[119]
- extend the date for rights holders to request "removal" of their works from the books database to April 5, 2011;[120]
- allow rights holders to direct the Registry to make their books available at no charge pursuant to one of several standard licenses (e.g., Creative Commons licenses) or similar contractual permissions for use authorized by the Registry;[121]
- allow the Registry, in its discretion, to authorize more than one free terminal per public library;[122] and

- specify that Google will not provide personally identifiable information about users to the Registry "other than as required by law or valid legal process."[123]

The revised settlement agreement prompts many of the same concerns that the October 2008 agreement prompted. Among other things, critics worry that the revised settlement still gives Google and/or the Registry effective control over orphan works; could have anticompetitive effects; fails to protect users' privacy; and does not adequately promote "fundamental library values."[124] However, while some commentators claim that the changes are "more cosmetic than substantial,"[125] others suggest that the changes might be substantive enough to "dampen large segments of the opposition" to the agreement and potentially increase the likelihood of its approval.[126]

The court granted preliminary approval of this revised agreement on November 19, 2009, and has scheduled a final fairness hearing on it for February 18, 2010.[127] Class members have until January 28, 2010, to opt into, opt out of, or object to the revised settlement.[128]

Potential Effects of Rejection or Approval of the Revised Settlement

Rejection of the revised settlement agreement could place the parties' claims and defenses back before the court. Moreover, even if eventually approved by the courts, the revised settlement agreement only governs claims against Google over its Library Project within the United States. Litigation in foreign jurisdictions remains possible. [129]

ACKNOWLEDGMENTS

Robin Jeweler, Legislative Attorney, original author of RS22356 *The Google Book Search Project: Is Online Indexing a Fair Use Under Copyright Law?*

End Notes

[1] Google Checks Out Library Books, Dec. 14, 2004, *available at* http://www.google.com/press/pressrel/print_library.html. Participating libraries included those at the University of Michigan, Harvard University, Stanford University, and Oxford University, as well as the New York Public Library.

[2] *Id.*

[3] *Id.* Copyright protection for books generally lasts "for a term consisting of the life of the author and 70 years after the author's death." 17 U.S.C. § 302(a).

[4] Google Checks Out Library Books, *supra* note 1.

[5] *Id.*

[6] Association of American University Presses, Google Book Search, Neé Google Print, *available at* http://www.aaupnet.org/aboutup/issues/gprint.html.

[7] Google Books Partner Program: Promote Your Books on Google—For Free, 2009, *available at* http://books.google.com/googlebooks/book_search_tour.

[8] *Id.*

[9] *See, e.g.*, Anandashankar Mazumdar, University Press Group Expresses Concern Over Google Print's Digitization of Works, 70 *Pat., Trademark & Copyright J.* 109 (June 3, 2005).

[10] 17 U.S.C. § 201(a). There are exceptions to this general rule, such as when a book is "made for hire" or is a "work of the United States Government." *See* 17 U.S.C. § 105 & 201(b).

[11] *See, e.g.*, Example Author Contract, *available at* http://www. writecontent. com/Publishing_ Tools/Author_ Contract _/ author_contract_.html ("The Author hereby grants to the Publisher exclusive rights to reproduce and/or publish or adapt and sell, and/or license third parties to publish or adapt and sell said Work.").

[12] 17 U.S.C. § 106(1) & (5).

[13] *See, e.g.*, Mazumdar, *supra* note 9.

[14] *See, e.g.*, Christine Mumford, Google Library Project Temporarily Halted to Allow Copyright Owner Response, 70 *Pat., Trademark & Copyright J.* 461 (Aug. 19, 2005).

[15] Authors Guild v. Google Inc., Class Action Complaint, No. 05 CV 8136 (S.D.N.Y. Sept. 20, 2005) at ¶ 20. The University of Michigan's library was the focus because Google began digitizing its books first. *Id.* at ¶ 31. Under copyright law, "literary works" are any "works, other than audiovisual works, expressed in words." 17 U.S.C. § 101.

[16] McGraw Hill Cos. v. Google Inc., Complaint, No. 05 CV 8881 (S.D.N.Y. Oct. 19, 2005). These companies were McGraw-Hill Companies; Pearson Education; Penguin Group; Simon & Schuster; and John Wiley and Sons.

[17] Fed. R. Civ. P. 23(e).

[18] *See, e.g.*, Susan Wojcicki, Google Print and the Authors Guild, Sept. 20, 2005, *available at* http://googleblog.blogspot.com/2005/09/google-print-and-authors-guild.html.

[19] *Id.*

[20] *See* 336 F.3d 811 (9th Cir. 2003). For more background on *Kelly*, see CRS Report RL33810, *Internet Search Engines: Copyright's "Fair Use" in Reproduction and Public Display Rights*, by Robin Jeweler and Brian T. Yeh.

[21] *See, e.g.*, Wojcicki, *supra* note 18.

[22] The Harvard University Libraries, for example, contain over 15 million books. *See* About the HOLLIS Catalog, June 25, 2007, *available at* http://lib.harvard.edu/catalogs/hollis.html.

[23] Anandashankar Mazumdar, Publishers: Value of Book Search Project Shows That Scanning Is Not Fair Use, 71 *Pat., Trademark & Copyright J.* 94 (Nov. 25. 2005).

[24] *Id.*

[25] P.L. 105-304. For more information on the DMCA generally, see CRS Report 98-943, *Digital Millennium Copyright Act, P.L. 105-304: Summary and Analysis*, by Dorothy M. Schrader.

[26] 11 U.S.C. § 512(b)-(c).

[27] *See* Field v. Google Inc., 412 F. Supp. 2d 1106 (D. Nev. 2006).

[28] *See, e.g.*, Class Action Complaint, *supra* note 15, at ¶ 23.d and ¶ 41 (alleging Google's infringement was willful); Complaint, *supra* note 16, at ¶ 2 (same).

[29] *See, e.g.*, Complaint, *supra* note 16, at ¶ 29 (arguing that Web pages differ from print books because rights holders in Web pages can rely on technological measures to prevent indexing, while authors of print books can take no such measures to prevent digitization).

[30] *See, e.g.*, Wojcicki, *supra* note 18.

[31] 17 U.S.C. § 107.

[32] *Id.*

[33] Campbell v. Acuff-Rose Music, Inc., 510 U.S. 569, 578 (1994).

[34] *Id.*

[35] Perfect 10, Inc. v. Amazon.com, Inc., 487 F.3d 701, 720 (9th Cir. 2007) (quoting the U.S. Constitution, art. I, § 8, cl. 8, as well as *Sony Corp. of Am. v. Universal City Studios, Inc.*, 464 U.S. 417, 429 n.10 (1984)).

[36] Stewart v. Abend, 495 U.S. 207, 237 (1990).

[37] Campbell, 510 U.S. at 577-78.

[38] *See, e.g.*, Sony, 464 U.S. at 451 ("Every commercial use of copyrighted materials is presumptively an unfair exploitation of the monopoly privilege that belongs to the owner of the copyright.").

[39] *See, e.g.*, Infinity Broad. Corp. v. Kirkwood, 150 F.3d 104, 108 (2d Cir. 1998) (retransmission of radio broadcast over telephone lines not a fair use); UMG Recordings, Inc. v. MP3.com, Inc., 92 F. Supp. 2d 349, 351 (S.D.N.Y. 2000) (reproducing analog audio CDs as MP3s not a fair use).

[40] Campbell, 510 U.S. at 578-79.

[41] *Id.* at 579.

[42] 336 F.3d 811 (9th Cir. 2003).

[43] *Id.* at 818-19.

[44] *Id.* at 818.

[45] *Id.* at 819.

[46] 487 F.3d 701, 721 (9th Cir. 2007), *rev'g* Perfect 10, Inc. v. Google Inc., 416 F. Supp. 2d 828 (C.D. Cal. 2006) (holding that Google's use of thumbnail versions of Perfect 10's copyrighted photographs was not fair, in part, because Google's thumbnails could potentially substitute for the reduced-size versions of these photographs that Perfect 10 had licensed another company to reproduce and distribute for display on cell phones).

[47] *Perfect 10, Inc.*, 487 F.3d at 721.

[48] *See, e.g., Kelly*, 336 F.3d at 816.

[49] Field, 412 F. Supp. 2d at 1115-16.

[50] A digital version of a print book would display poorer resolution than the original. However, it would enable researchers to locate specific content more easily by using the "search" or "find" functions of their Web browsers.

[51] 464 U.S. 417 (1984).

[52] *Id.* at 442.

[53] *See, e.g.*, Jonathan Band, The Google Print Library Project: Fair or Foul?, 9 *J. of Internet L.* 1, 4 (Oct. 2005); Christopher Heun, Courts Unlikely to Stop Google Book Copying, *InternetWeek* (Sept. 2, 2005), *available at* http://internetweek.cmp.com/showArticle.jhtml?articleID=170700329.

[54] 92 F. Supp. 2d 349, 352 (S.D.N.Y. 2000) ("[D]efendant's 'fair use' defense is indefensible and must be denied as a matter of law.").

[55] *Id.*

[56] *Kelly*, 336 F.3d at 820.

[57] *See, e.g.*, Feist Publ'ns, Inc. v. Rural Tel. Serv., 499 U.S. 340 (1991) (requiring originality in the selection or arrangement of facts for copyrightability).

[58] *See, e.g.*, Baker v. Selden, 101 U.S. 99 (1879) (distinguishing non-protectable ideas from their protectable expressions).

[59] *Campbell*, 510 U.S. at 577-78.

[60] Worldwide Church of God v. Philadelphia Church of God, Inc., 227 F.3d 1110, 1118 (9th Cir. 2000) (internal quotations omitted).

[61] *See, e.g., Kelly*, 336 F.3d at 821; Sega Enters., Ltd. v. Accolade, Inc., 977 F.2d 1510, 1523 (9th Cir. 1992).

[62] *See, e.g.*, Harper & Row Publishers, Inc. v. Nation Enters., 471 U.S. 539, 564 (1985).

[63] *See, e.g., Kelly*, 336 F.3d 811; *Perfect 10, Inc.*, 487 F.3d 701.

[64] *See Field*, 412 F. Supp. 2d at 1115-16.

[65] *See Kelly*, 336 F.3d at 816.

[66] *See* Complaint, *supra* note 16, at ¶ 5 (noting that publishers were already making their print books available online in various ways, including a partnership with the search engine Yahoo!).

[67] *Cf.* BMG Music v. Gonzalez, 430 F.3d 888, 891 (7th Cir. 2005) ("Copyright law lets authors make their own decisions about how best to promote their works.").

[68] *See, e.g.*, Bill Graham Archives, LLC v. Dorling Kindersley, Ltd., 386 F. Supp. 2d 324, 332 (S.D.N.Y. 2005).

[69] *Perfect 10, Inc.*, 487 F.3d at 725.

[70] *See, e.g., Campbell*, 510 U.S. at 591.

[71] Fed. R. Civ. P. 23(e). The court may reject a proposed settlement because of concerns not raised by class members or other parties to the agreement. *See, e.g.,* Muchnick v. Thompson Corp., 509 F.3d 116 (2d Cir. 2007) (quashing the proposed settlement agreement resolving the litigation in *Tasini v. New York Times* because some members of the proposed settlement class had not registered their works with the U.S. Copyright Office and so lacked standing to bring suit in federal court). The district court had previously approved the settlement, and no class member or party had raised this objection.

[72] Authors Guild, Inc. v. Google Inc., Settlement Agreement, Case No. 05 CV 8136-JES (S.D.N.Y. Oct. 28, 2008).

[73] *Id.* at ¶ 2.1(a) (providing that Google would pay 70% of the net revenue earned from uses of Google Book Search in the United States to rights holders); ¶ 2.1(b) (providing that Google would pay at least $45 million into a "Settlement Fund," whose proceeds would pay rights holders whose books or "inserts" were digitized prior to January 5, 2009).

[74] *Id.* at ¶ 2.1(c). Among other functions, the Registry could negotiate the terms of "New Revenue Models" (e.g., print-on-demand) with Google and negotiate pricing categories and percentages for sale of digitized materials to users.

[75] Because this license is non-exclusive, the Registry could license other entities to digitize, index, or display the works of rights holders. However, if the Registry were to enter into a similar agreement within 10 years of the settlement's effective date, it must extend comparable economic and other terms to Google. *Id.* at ¶ 3.8(a).

[76] *Id.* at ¶ 3.1. Google's rights to use books within their terms of copyright protection would hinge upon whether they were "commercially available," or available "for sale new through one or more then-customary channels of trade in the United States." *See id.* at ¶ 1.28. If a book is commercially available, Google could not make "display uses" without the copyright holders' consent. *Id.*azzzz at ¶¶ 3.3-3.5. Conversely, if a book is not commercially available, Google could make "display uses" unless the rights holder objects. *Id.* This distinction between commercially available and non-commercially available books would significantly vary the legal protections of copyright law, which protects all works equally, regardless of their commercial availability, during their terms of copyright protection. *See* 17 U.S.C. § 106 and § 302.

[77] *See, e.g.*, Settlement Agreement, *supra* note 72, at ¶ 3.7.

[78] *Id.* at ¶ 7.2(f)(i)-(ii) and Article X.

[79] *See, e.g.*, James Grimmelmann, Principles and Recommendations for the Google Book Search Settlement, Nov. 8, 2008, *available at* http://www.laboratorium.net/archive/2008/11/08/ principles_ and_ recommendations _for_ the_ google book (noting the existence, as well as the potential inadequacy, of provisions regarding users).

[80] Settlement Agreement, *supra* note 72, at ¶1.49.

[81] Authors Guild, Inc. v. Google Inc., Order Granting Preliminary Settlement Approval, Case No. 05 CV 8136-JES (S.D.N.Y. Nov. 17, 2008).

[82] *See, e.g.*, Court Delays Google Book Search Settlement Hearing, Extends Opt-Out Period, 78 *Pat., Trademark & Copyright J.* 9 (May 1, 2009).

[83] *See, e.g.*, Anandashankar Mazumdar, DOJ Urges Rejection of Google Settlement; Plaintiffs Move to Adjourn Final Hearing, 78 *Pat., Trademark & Copyright J.* 631 (Sept. 25, 2009).

[84] *See, e.g.*, James Gibson, Google's New Monopoly? How the Company Could Gain by Paying Millions in Copyright Fees, *Wash. Post*, Nov. 3, 2008, at A21 ("[S]ettling probably puts Google in a better position than it would have been in if it had won its case in court. ... Google's concession has made it more difficult for *anyone* to invoke fair use for book searches [and] [b]y settling the case, Google has made it much more difficult for others to compete with its Book Search service.").

[85] *See, e.g.*, Settlement Agreement, *supra* note 72, at ¶ 3.1(a).

[86] *See, e.g.*, Miguel Helft, Some Raise Alarms as Google Resurrects Out-of-Print Books, *New York Times*, Apr. 4, 2009, at A1 ("Nothing prevent[s] a potential rival from following in [Google's] footsteps—namely, by scanning books without explicit permission, waiting to be sued and working to secure a similar settlement."). Microsoft, for example, has explored offering book-search capabilities in the past. *See* Andrea L. Foster, Microsoft's Book-Search Project Has a Surprise Ending, *Chron. of Higher Educ.*, May 29, 2008, *available at* http://chronicle.com/free/2008/05/ 3022n.htm.

[87] *See, e.g.*, Settlement Agreement, *supra* note 72, at ¶ 3.8(a) ("The Registry ... will extend economic and other terms to Google that, when taken as a whole, do not disfavor or disadvantage Google as compared to any other substantially similar authorizations granted to third parties by the Registry.").

[88] *See, e.g.*, Laura G. Mirviss, Harvard-Google Online Book Deal at Risk, *The Harvard Crimson*, Oct. 30, 2008, *available at* http://www.thecrimson.com/article.aspx?ref=524989 ("The settlement provides no assurance that the prices charged for access will be reasonable.").

[89] *See, e.g.*, Letter to the Honorable Denny Chin, Re. The Authors Guild Inc., et al. v. Google, Inc., Civil No. 1:05-CV-8136, *available at* http://pub.bna.com/ptcj/DOJLetterEtc.pdf; Miguel Helft, U.S. Inquiry Is Confirmed into Google Books Deal, *N.Y. Times*, July 2, 2009, *available at* http://www.nytimes.com/2009/07/03/technology/companies/ 03google.html?_r=1&ref=technology.

[90] Nathan Pollard, Opposition Group to Google Book Settlement Includes Big Names, Authors and Libraries, 78 *Pat., Trademark & Copyright J.* 539 (August 28, 2009).

[91] *See, e.g.*, Anandashankar Mazumdar, Internet Archive Warns of Online Book Market Consolidation from Google Settlement, *Pat., Trademark & Copyright L. Daily*, Apr. 23, 2009. For more on orphan works generally, see CRS Report RL33392, *"Orphan Works" in Copyright Law*, by Brian T. Yeh.

[92] Grimmelmann, *supra* note 79.

[93] *See, e.g.*, Mazumdar, *supra* note 91.

[94] *Id.*

[95] *See, e.g.*, Anandashankar Mazumdar, Google Books Settlement Prompts Questions about Effect on Readers, Libraries, Others, *Pat., Trademark & Copyright L. Daily*, Mar. 18, 2009 (quoting Marybeth Peters, Register of Copyrights).

[96] Orphan Works Act of 2008, H.R. 5889, 110th Cong. There was a similar bill in the Senate, but it did not include the notice requirements of the House bill. *See* Shawn Bentley Orphan Works Act of 2008, S. 2913, 110th Cong.

[97] *See, e.g.*, Helft, *supra* note 86 ("They are doing an end run around the legislative process.") (quoting Brewster Kahle, founder of the Internet Archive and the Open Content Alliance).

[98] American Library Association, Association of Research Libraries, and Association of College and Research Libraries, Google Library Project Settlement, July 29, 2009, *available at* http://librarian.lishost.org/?p=2731.

[99] Nathan Pollard, Urban Libraries Council Suggests Revisions to Google Book Search Settlement, 78 *Pat., Trademark & Copyright J.* 540 (August 28, 2008).

[100] Center for Democracy and Technology, Privacy Recommendations for the Google Book Search Settlement, July 27, 2009, *available at* http://cdt.org/copyright/20090727_GoogleRecs.pdf. CDT filed an amicus brief with the court on September 4, 2009, expressing similar concerns. *See* Authors Guild v. Google, Inc.: Brief Amicus Curiae of the Center for Democracy and Technology in Support of the Approval of the Settlement and Protection of Reader Privacy, *available at* http://www.cdt.org/copyright/CDT-GoogleAmicusFinal.pdf

[101] Miguel Helft & Motoko Rich, Lawyer and Author Adds His Objections to Settling the Google Book Lawsuit, *New York Times*, Aug. 18, 2009, *available at* http://www.nytimes.com/2009/08/19/technology/internet/19google.html. Some commentators have described this challenge as the "most fundamental challenge to the settlement yet." *See id.* (quoting James Grimmelmann, associate professor at the Institute for Information Law and Policy at New York Law School).

[102] *Id.*

[103] *See, e.g.*, National Writers Union Opposes Settlement of Google Books Dispute, 78 *Pat., Trademark & Copyright J.* 515 (Aug. 21, 2009).

[104] *Id.*

[105] *Id.*

[106] *See, e.g.*, Nathan Pollard & Christine You, Benefits of Google Book Deal Praised, But Orphan Works Impact Still in Doubt, 14 *Electronic Commerce & Law Report* 1184 (Aug. 19, 2009).

[107] *See, e.g.*, Josh Hadro, Michigan Deal a New Twist on Access to Scanned Book Content: The Effort Marks Another Repurposing of Digital Editions Scanned In-house and by Google, *Library Journal*, July 23, 2009, *available at* http://www.libraryjournal.com/article/CA6672693.html.

[108] *See, e.g.*, Nathan Pollard, Social Justice IP Panel Says Google Book Settlement Will "Level the Playing Field," 78 *Pat., Trademark & Copyright J.* 408 (July 31, 2009).

[109] *See, e.g.*, Pollard & You, *supra* note 106 (quoting Ed Black, President of the Computer & Communications Industry Association).

[110] *See, e.g.*, Pollard, *supra* note 108.

[111] *Id.* (quoting Charles S. Brown, advisor to the president, National Association of the Blind).

[112] *Id.* (quoting Lateef Mtima, Professor of Law, Howard University; Wade Henderson, President, Leadership Conference on Civil Rights; and Brent A. Wilkes, National Executive Director, League of United Latin American Citizens).

[113] *See, e.g.*, Joe Kirwin, EU Commissioner Endorses Google Digital Book Settlement with U.S. Publishers, *Pat., Trademark & Copyright L. Daily*, Aug. 31, 2009.

[114] *Id.*

[115] *See* Authors Guild, Inc. v. Google Inc., Revised Settlement Agreement, Case No. 05 CV 8136-JES (S.D.N.Y. Nov. 13, 2008).

[116] *Id.* at ¶ 1.19. A published work is a "United States work," as defined under the Copyright Act, only when it is (1) first published in the United States; (2) simultaneously published in the United States and another treaty party whose law grants a term of copyright protection that is the same as or longer than the term provided in the United States; (3) simultaneously published in the United States and a foreign nation that is not a treaty party; or (4) published in a foreign nation that is not a treaty party and all the authors of the work are nationals, domiciliaries, habitual residents, or, in the case of audiovisual works, headquartered in the United States. *See* 17 U.S.C. § 101.

[117] Additionally, Google will be able to take advantage of any future legislative changes, such as legislation allowing use of orphan works, that puts Google at a competitive disadvantage only if a "third party is actually taking advantage of such law(s) in connection with services that competitively disadvantage Google in its provision of any such Google Products and Services." Revised Settlement Agreement, *supra* note 115, at ¶ 3.8.

[118] *Id.* at ¶ 2.4.

[119] *Id.* at ¶ 6.2(b)(iii).

[120] *Id.* at ¶ 1.126(a). Provided the request is made by April 5, 2011, digital copies of the book will not be "accessible" to Google or participating libraries, although they may continue to exist on back-up tapes or other electronic back-up storage media. If the request is made after April 5, 2011, but before March 9, 2012, digital copies of the book will not be "accessible" to Google, but Google may use the copies to provide digital copies and make ongoing technological updates for fully participating libraries and host sites and when acting as a host site as authorized under the revised agreement. *Id.* at ¶ 1.126(b). When a work is "accessible" to Google, it generally may be displayed in certain ways. *See id.* at ¶ 1.1.

[121] *Id.* at ¶ 4.2(a)(i).

[122] *Id.* at ¶ 4.8(a)(i)(3).

[123] *Id.* at ¶ 6.6(f).

[124] *See, e.g.*, Norman Oder, Revised Google Settlement Offers Minor Changes on Antitrust Issue, No Response on Library Pricing, *Library Journal*, Nov. 14, 2009, *available at* http://www.libraryjournal.com/article/CA6707181.html? desc=topstory; John Timmer, Google Book Settlement Revised, Criticized, *Ars Technica*, Nov. 16, 2009, *available at* http://arstechnica.com/media.

[125] Timmer, *supra* note 124.

[126] Anandashankar Mazumdar, Revised Google Book Search Settlement Proposals Removes Foreign Works from Mix, *Pat., Trademark, & Copyright L. Daily*, Nov. 17, 2009 (quoting James Grimmelmann, Associate Professor at New York Law School).

[127] Authors Guild, Inc. v. Google Inc., Preliminary Approval of Proposed Settlement Agreement, Case No. 05 CV 8136(DC) (S.D.N.Y. Nov. 19, 2009).

[128] *Id.*

[129] *See, e.g.*, Editions du Seuil v. Google Inc., Tribunal de Grand Instance de Paris (alleging that Google Book Search infringes copyright under French law); Joe Kirwin, EU States Urge Investigation into Google's Digital Book Plan, *Pat., Trademark & Copyright L. Daily*, May 29, 2009 (noting that the Federation of European Publishers has also raised concerns about the potential anticompetitive effects of the Google Library Project with the European Commission).

In: Digital Books: Competition and Commerce
Editor: Oliver A. Hagen
ISBN: 978-1-61728-299-7

Chapter 2

"ORPHAN WORKS" IN COPYRIGHT LAW

Brian T. Yeh

SUMMARY

Orphan works are copyrighted works whose owners are difficult or impossible to identify and/or locate. Orphan works are perceived to be inaccessible because of the risk of infringement liability that a user might incur if and when a copyright owner subsequently appears. Consequently, many works that are, in fact, abandoned by owners are withheld from public view and circulation because of uncertainty about the owner and the risk of liability.

In 2006, at the request of Congress, the U.S. Copyright Office issued its *Report on Orphan Works* ("Report"). The goal of the Report was to elicit public comment and evaluate the extent of real or perceived problems that content users encounter in their efforts to use these works. The Report defines the problems it identified, and concludes that the problem is indeed real and should be addressed legislatively. It analyzes stakeholders' views on the issue and constraints on solutions imposed by the structure of U.S. copyright law and international copyright obligations. The Report sets forth a proposal to amend the Copyright Act by adding a provision that would limit liability for infringing use of orphan works when, prior to use, a user performs a reasonably diligent search for the copyright owner and provides attribution to the author and copyright owner, if possible. In some instances, when copyright infringement is made without commercial advantage and the user ceases infringement promptly after receiving notice thereof, no monetary relief would be available.

Adopting many of the suggestions of the Copyright Office, the Orphan Works Act of 2006 was introduced in the 109th Congress, second session (H.R. 5439). This bill was later incorporated into an omnibus copyright bill, appearing as Title II of The Copyright Modernization Act of 2006 (H.R. 6052). However, the bill was not addressed by the end of that Congress's adjournment. The bill would have implemented a limitation on monetary damage liability for specified infringement of orphan works, but took a more detailed approach than the Report's original proposals in establishing requirements for such liability limitations, such as articulating standards for a "reasonably diligent search." The bill would also have directed the Copyright Office to study and report on the implementation of the new orphan works amendment, and to study and make recommendations for a "small claims" procedure to address copyright infringement.

Legislation addressing the orphan works issue was reintroduced in the 110th Congress: the Orphan Works Act of 2008 (H.R. 5889) and the Shawn Bentley Orphan Works Act of 2008 (S. 2913). The two bills resembled the Orphan Works Act of 2006, although there were substantial differences from that earlier legislation and even between themselves. These additional or revised provisions were added in part to address concerns raised by photographers, illustrators, and other visual artists, as well as textile and home furnishing manufacturers. While S. 2913 passed the Senate, H.R. 5889 did not make it out of the House Judiciary Committee.

This chapter surveys the findings and conclusions in the U.S. Copyright Office's *Report on Orphan Works* and analyzes the orphan works bills that were considered by the 109th and 110th Congresses. No legislation relating to orphan works has yet been introduced in the 111th Congress as of the date of this chapter. However, the outcome of the Google Book Search class action lawsuit (and its pending settlement) may potentially affect future orphan works legislation.

BACKGROUND

In response to requests by several Members of Congress in January 2005, the U.S. Copyright Office agreed to examine issues surrounding "orphan works." Orphan works are copyrighted works whose owners are difficult or impossible to identify and/or locate. The Copyright Office issued a Notice of Inquiry requesting public comment from interested parties on the subject.[1] The Office accepted written comments and hosted public roundtable discussions on the topic. In January 2006, it issued its Report on Orphan Works, which includes proposed legislative language to address the problem identified.[2] Hearings were held in the 109th and 110th Congresses on the orphan works problem.[3]

This chapter surveys the findings of the Report on Orphan Works ("Report"), considers the Copyright Office's proposed amendment to the Copyright Act to address the issue, and analyzes introduced orphan works legislation in the 109th and 110th Congresses. No similar legislation has yet been introduced in the 111th Congress.

Defining the Problems Associated with Orphan Works

The constitutionally authorized grant of a limited monopoly to copyright holders is intended "To promote the Progress of Science and useful Arts" by producing incentives for creative works and their dissemination to the public.[4] Ultimately, it is the public interest that supports allowing copyright holders to financially exploit the value of their creative efforts by controlling access to protected work. Someone who wants to exercise one or more of the copyright holder's exclusive rights in a copyrighted work must obtain permission to do so.[5] The terms for usage and recompense, if any, are negotiated and agreed to by the rights' holder and the prospective user.[6]

When an owner cannot be identified or located, a protected work is an "orphan" work. Many believe that orphan work status renders a work inaccessible. The inaccessibility arises from the risk of liability that a user might incur for copyright infringement if and when a copyright owner subsequently appears:

> First, the economic incentive to create may be undermined by the imposition of additional costs on subsequent creators wishing to use material from existing works. Subsequent creators may be dissuaded from creating new works incorporating existing works for which the owner cannot be found because they cannot afford the risk of potential liability or even of litigation. Second, the public interest may be harmed when works cannot be made available to the public due to uncertainty over its copyright ownership and status, even when there is no longer any living person or legal entity claiming ownership of the copyright or the owner no longer has any objection to such use.[7]

This risk of infringement may be particularly burdensome when a creator incorporates protected work into a new adaptation or transformative work. How then is public policy best served by facilitating the public's access to and use of such a work? And how best to define what constitutes an orphan work for infringement purposes, to facilitate access to orphan works, and to promote their use without vitiating the copyright or unfairly appropriating the work into the public domain?

Obstacles to Obtaining Permission

Copyright law is the engine driving a vast private market of rights' holders and users. The structure of the law in many ways shapes the intellectual property (IP) marketplace for negotiations between owners and users, but it does not control all aspects of it. Notifying the public of ownership is the responsibility of the rights' holder. Determining whether a work is protected and identifying the actual owner of the copyright (who may or may not be the creator) is the responsibility of the prospective user. But the identification process can be extremely complicated, difficult, and in many cases, prohibitively costly. There are many components to the determination of whether something is likely to be covered by copyright.[8] The prospective user must first make a preliminary determination as to whether a work is indeed copyrighted or has passed into the public domain.[9] Changes to the term of copyright effected by repeal of the 1909 law and adoption of the 1976 Act, subsequent extensions to the term, and the abandonment of "formalities" (discussed *infra*), all work to complicate calculations of the likely subsistence of copyright, particularly with respect to works created prior to 1978.[10]

A Universal Registry of Copyright Owners

Although registration with the U.S. Copyright Office is most authoritative, there is no universal copyright registry. Various registries or databases exist to allow identification of copyright holders in various industries or mediums, but they are essentially voluntary, so checking with a database may not be dispositive regarding copyright status and/or ownership.[11] Furthermore, because IP is indeed property, through sale, assignment, or bequest, over time, ownership rights may be transferred. Older works of minimal commercial value may essentially be neglected or abandoned. Finding a copyright owner for them can be challenging.

Formalities

Under the 1909 Copyright law, there were many specific actions, i.e., "formalities," that needed to be taken by the creator/owner in order to create a valid copyright. Failure to do so could void the copyright. Among the essential formalities were posting of a notice of copyright on a work and registration with the U.S. Copyright Office. At the expiration of the first 28-year term of copyright, a renewal had to be filed to extend protection for another 28-year term.[12] Observing formalities as a prerequisite to creating a valid copyright was abandoned under the 1976 Copyright Act. Under current law, a copyright is created automatically when the creative expression is fixed in tangible form. Copyright formalities were rejected in the 1976 law for several reasons. The legislative history notes the concern that rigid formalities put an undue burden on creators, who could lose copyright protection in its entirety for failure to comply with a formality requirement.[13] A primary goal, however, was to harmonize U.S. copyright law with international treaties and practice, where formalities are not a requirement for copyright protection.

Nevertheless, changes to U.S. law significantly complicated the process of identifying copyright holders. One consequence of the formalities requirements associated with copyright creation was notice and registration. A search of copyright registration records was more—though not definitively—likely to help a prospective user determine both copyright status and owner information.

Copyright Infringement Litigation and Damages

Under the current law for works created after 1978, an owner *may* register a work at any time during the subsistence of the copyright.[14] A work *must* be registered prior to the rights' holder bringing suit for infringement;[15] registration is also necessary in order for a owner to seek statutory damages for infringement.[16]

In the event that a court finds copyright infringement, it may issue an injunction to prevent or stop it,[17] and award monetary damages. Damages may be the actual value of lost profits, or damages set by statute, known as "statutory damages."[18] Statutory damages prescribe amounts that may be significantly higher than actual damages for lost profits—from $750 to $150,000. The amount of statutory damages may be increased in cases where a court finds that infringement was willful or, correspondingly, reduced when it finds the infringement was "innocent," i.e., the infringer was infringer had reasonable grounds to believe that the use was a fair use under § 107. A court may also award court costs and attorneys' fees.[19] In other words, registration, with its effect of creating a searchable record and thereby providing public notice of ownership, is not legally required to *create* a copyright, but to *enforce* it. The existence of statutory damages and the award of attorneys' fees facilitates enforcement of infringement liability by rights' holders when actual damages may not support the costs of litigation.

THE REPORT ON ORPHAN WORKS

By conducting stakeholder discussions and reviewing extensive submissions of comments, the U.S. Copyright Office's study considers the landscape surrounding orphan works.

At the outset, it sets forth what were *not* considered to be orphan work problems, namely, situations where a prospective user contacted the owner but did not receive permission to use the work.[20] The analysis also narrows the situations in which it views orphan works as presenting an insurmountable problem to prospective users. It delineates several provisions of the copyright law that might permit use of an orphan work (or any copyrighted work) absent an owner's permission:

- The "idea/expression" dichotomy, rooted in the First Amendment and codified at 17 U.S.C. 102(b), prohibits copyright protection for ideas, procedures, concepts, etc. that may otherwise be embodied in a copyright-protected work.[21] This jurisdictional limitation on copyright protection may be especially useful to prospective users of works of non-fiction, and "utilitarian" works like computer programs, textbooks, manuals, etc.[22]
- Fair use, codified at 17 U.S.C. § 107, permits limited use of copyright-protected work for purposes such as criticism, comment, news reporting, teaching, scholarship, or research.
- Other express exemptions in the Copyright Act at § § 108, 110, and 117 allow specified uses of copyrighted works associated with preservation, education, and religious activities.

The Report identifies many obstacles to identifying and locating copyright owners and assigns general categories of uses that appear to be most impacted by orphan works, namely, uses by "subsequent creators" who may create a derivative commercial work incorporating the orphan work; "large scale access uses" by institutions such as libraries that make available a wide body of work to the public; "enthusiast" uses by individuals who have an interest in a particular work, subject, or artist; and "private" uses, the most common illustration being someone who wishes to reproduce a family photograph or make a potentially infringing use of obsolete or orphaned computer software.[23]

The Report explains that the 1976 Copyright Act arguably exacerbated the orphan works problem by abandoning formalities such as renewal registration, and why the international copyright regime to which the United States is a signatory both precludes a re-adoption of formalities and limits the scope of permissible exemptions to the copyright holders' rights.[24]

The study reviews solutions proposed by those involved in the orphan works dialogue. It groups and considers them in four categories, described in the Report as follows:

- *Solutions that already exist under current law and practice.* These were usually noted only in passing; commenters (even commenters opposed to any orphan works provision) did not take the position that the existing law is sufficient to solve the orphan works problem.

- *Non-legislative solutions.* An example of a solution in this category is a proposal for improved databases for locating owners of works. These solutions were also usually noted only in passing, and were not advanced as sufficient to fix the problem.

- *Legislative solutions that involve a limitation on remedies when a user uses an orphan work.* The most substantive comments fell into this category, and most of the comments by professional organizations or academics fell into this category.

- *Other legislative solutions.* Examples of proposed solutions in this category are deeming all orphaned works to be in the public domain, or changing the tax or bankruptcy codes to reduce the factors that cause orphan works to come into existence in the first place.[25]

It also considers several of the solutions proposed. For example, one approach might be that utilized by the Canadian Copyright Board, which reviews applications for use of orphan works and approves them prior to use. This method receives support by some for the certainty that it provides and opposition by others who view it as administratively cumbersome, expensive, and largely ineffective in promoting actual use of orphan works.

The Copyright Office's Recommendation

The Report concludes that the orphan works problem, though difficult to describe and quantify, is indeed real. Though some instances of non-infringing use of such works may be effected under other sections of the law, there are still many situations in which prospective users lack guidance on whether and how they may use orphan works, and authority to do so. The Report recommends statutory language to remedy the orphan works problem,[26] with a detailed supporting rationale.

The proposal takes the approach of limiting remedies for the copyright owner if a user satisfies new statutory requirements for use of an orphan work. The proposed language would add a new § 514 under chapter 5 of the Copyright Act, dealing with copyright infringement and remedies. One who uses an orphan work would be required to have performed "a good faith, reasonably diligent search" to identify the copyright holder and provide "attribution to the author and copyright owner of the work, if possible and appropriate." If the user of a orphan work who has satisfied the search and attribution requirements is subsequently sued by the rights' holder for infringement, the owner would be limited to "reasonable compensation for the use of the infringed work." When the infringement is made without commercial advantage and the user ceases infringement promptly after receiving notice thereof, no monetary relief would be available.

Injunctive relief, i.e., prohibiting continuing use of the infringing work, would *not* be available when the orphan work is incorporated into a derivative work that uses the protected work in a transformative manner, provided that the infringer pays reasonable compensation to the copyright owner and provides attribution to the protected work as reasonable. In all other cases, the court may impose injunctive relief to prevent the continuing infringement, but would be directed to consider the harm that relief would cause the infringer who has complied with orphan works requirements in making the infringing use.

The proposed language specifies that nothing in its provisions would affect other rights, limitations or defenses to copyright infringement, including fair use. The provision would sunset ten years after enactment.

The goal of the proposal is to promote the good-faith use of true orphan works by limiting damages available in the event that an owner appears and the user is subsequently charged with infringement. The proposed solution attempts to balance several competing interests:

Notice

Some content users complain about the lack of easy-to-use comprehensive sources of information identifying copyright owners. Easier access to ownership information would minimize mistakes by users in calculating whether a work is actually an orphan work. But a solution that imposes notice or identification requirements on rights' holders as a condition of protecting their copyright would violate both the Copyright Act and international treaty obligations if its effect was to reinstate formality requirements. While it is obviously in the interest of copyright holders to make the public aware of ownership, the proposal would not impose additional regulatory burdens on owners, or the government, by establishing new reporting mechanisms.

Certainty versus Flexibility

Many who promote access to orphan works seek a system that best assures potential users that they will be exempt from copyright infringement liability *prior* to usage. But any proposed orphan work exemption will potentially affect a vast array of industries and media, such as movies, music, books, and photographs. There are different physical characteristics, traditions, standards, and business practices which affect the ease of researching ownership and obtaining permissions for any given medium. Likewise, different users have different goals, such as nonprofit versus commercial usage.

The approach suggested is in many ways comparable to copyright's well-known "fair use" exemption in its breadth and flexibility.[27] Like fair use, the orphan work exemption would be a *defense* to copyright infringement. The proposal takes a case-by-case approach that would give a court discretion to consider behavior by both the user and claimant. Did the user perform a "reasonably diligent search" with proper attribution? Did the claimant decline to accept "reasonable compensation" for the identified infringement, which, under the proposal, becomes, in effect, a statutory cap on relief available? Arguably, it would share many of the strengths and weaknesses of fair use. Among the former is flexibility to accommodate a broad range of media and situations. Among the latter may be difficulty assessing the likelihood of the success of the defense, and costs that may be unintentionally incurred.

Standards

The proposal does not define terms such as "reasonably diligent search," although much discussion is provided. Best practices for media-specific searches are likely to evolve over time through collaborative efforts and judicial interpretation. Likewise the notion of reasonable compensation is a fluid one, another factor that is viewed as advantageous or non-advantageous by different parties. Critics among users point to difficulties when the amount of liability exposure is uncertain. Critics among owners worry that courts interpreting the term may depress the value of "reasonable compensation," by valuing it at what the user proposes to pay absent negotiations. They fear that it may amount to a statutory royalty rate.

Damages

Because the proffered exemption is a defense to copyright infringement, the costs of litigation were considered in the discussion. Indeed, the Report spells out at great length the concerns expressed by both content owners and users on the burdens imposed by having to litigate a claim of or a defense to infringement. Users argue that the prospect of statutory damages has a chilling effect on their use of valuable historic material, for example, documentary film footage. But many owners assert that a limitation on the remedies for infringement would make enforcement impracticable. They simply cannot enforce their copyright if the enforcement costs more than recoverable damages.

Visual Arts

Photography and visual arts pose special challenges for copyright ownership identification generally, and, consequently, in connection with orphan works. By their very nature, they are difficult to source. Critics are concerned that the orphan work proposal would affect illustrations and photographs disproportionately because images are commonly published, by tradition or business practice, without identifying information. If a visual representation contains identifying information, it may be, and often is, easily removed. Verbal registries cannot adequately describe visual representations, e.g., "nine abstract dogs in an abstract garden."[28] Visual registries may contain prohibitively voluminous entries and be too difficult to search. They fear enactment of the orphan works proposal might interfere with commercial markets for visual work; that it could have the effect of "legalizing" infringement where ever the rights' holder cannot be identified or located; that it will put too great a burden on rights' holders to exercise diligence in monitoring infringing use; and that limiting recoverable damages will make enforcement actions economically unfeasible. The *de facto* result, they contend, would deprive visual artists of meaningful copyright protection.[29]

Legislative Proposals

No legislation relating to orphan works has yet been introduced in the 111th Congress as of the date of this chapter. What follows is an analysis of orphan works bills that were considered by the 109th and 110th Congresses.

H.R. 5439, 109th Congress, Second Session, the Orphan Works Act of 2006

The Orphan Works Act of 2006 incorporated many of the recommendations of the Copyright Office and was introduced and reported by the House Subcommittee on Courts, the Internet, and Intellectual Property in May 2006. This bill was later imbedded in an omnibus copyright bill, appearing as Title II of The Copyright Modernization Act of 2006 (H.R. 6052). The bill provided significantly greater detail than the Copyright Office's proposed language, including setting forth specific standards to establish what is a "reasonably diligent search." However, no orphan works legislation was passed by the end of the 109th Congress' adjournment.

H.R. 5889, 110th Congress, Second Session, the Orphan Works Act of 2008

The Orphan Works Act of 2008 resembled the 109th Congress's orphan works legislation although it had substantial differences. The bill would have added a new § 514 to the Copyright Act entitled "Limitation on remedies in cases involving orphan works." It essentially would have implemented the Copyright Office's proposal to limit liability for an infringing use of an orphan work. As a prerequisite to qualifying for the limitation, the infringer would have had to satisfy several conditions prior to using the orphan work, including performing and documenting a "qualifying search" in good faith to locate the owner of the infringed copyright and filing with the Register of Copyrights a "Notice of Use."[30] H.R. 5889 provided that a search is qualifying if the infringer undertakes a "diligent effort" to locate the owner of the infringed copyright.[31]

H.R. 5889 required the Copyright Office to establish and maintain an archive that retained the "Notice of Use" filings that would have been submitted by the infringer prior to using the copyrighted work. H.R. 5889 also directed the Register of Copyrights to undertake a "certification process" for the creation of electronic databases that facilitate the search for pictorial, graphic, and sculptural works.[32]

Limitations on Remedies

The heart of H.R. 5889 was the limitation on monetary relief that may be awarded by a court (including actual damages, statutory damages, costs, and attorney's fees) when the use of an orphan work is found to be infringing. Users who had satisfied statutory criteria would have been required to pay "reasonable compensation" for the use of the infringed work.[33] H.R. 5889 defined "reasonable compensation" to mean "the amount on which a willing buyer and willing seller in the positions of the infringer and the owner of the infringed copyright would have agreed with respect to the infringing use of the work immediately before the infringement began."[34] The bill also permitted a court to consider, in determining the amount of reasonable compensation, the value (if any) that has been added to a work due to the fact that the work is registered with the Copyright Office.[35]

Safe Harbor

H.R. 5889 exempted certain infringers of orphan works from the requirement to pay reasonable compensation for the use of infringed works. This statutory "safe harbor" would have been available if the infringer is a nonprofit educational institution, library, or archives, or a public broadcasting entity, *and* the infringer proved by a preponderance of the evidence that:

- The infringement was performed without any purpose of direct or indirect commercial advantage and for a charitable, religious, or educational purpose; and
- The infringer promptly ceased the infringing use after receiving notice of the claim for infringement and after conducting an expeditious good faith investigation of the claim.[36]

Exceptions to the Eligibility for Limitation on Monetary Remedies

H.R. 5889 denied the limitation on remedies for infringers who, after receiving a notice of the claim for infringement[37] and having a chance to conduct an "expeditious" good faith investigation of the claim, either: (1) fails to negotiate in good faith "reasonable compensation" with the copyright owner; or (2) fails to render payment of reasonable compensation in a reasonably timely manner.[38] In addition, the bill noted that failure to comply with any of the eligibility requirements for the limitation on remedies would have subjected the infringer to all available remedies for civil copyright infringement.[39]

Injunctive Relief

The bill permitted a court to award injunctive relief to prevent or restrain any infringement. H.R. 5889 provided an exception to this general rule, however: a court may *not* enjoin the infringing use of an orphan work when it is incorporated (or starting to be integrated) into a new work of authorship, so long as the infringer pays reasonable compensation and provides attribution to the owner of the infringed work in a manner that is reasonable under the circumstances, if requested by such owner.[40]

Exclusion for Useful Articles

H.R. 5889 provided that the limitations on monetary and injunctive relief were *unavailable* to "an infringer for infringements resulting from fixation of a work in or on a useful article that is offered for sale or other distribution to the public."[41] The Copyright Act defines "useful article" to mean "an article having an intrinsic utilitarian function that is not merely to portray the appearance of the article or to convey information."[42] Examples of useful articles include shower curtains, textile designs, wall coverings, home furnishings, coffee mugs, and clothing with images on them. Therefore, infringers of such useful articles are not allowed to assert the right to claim the limitation on the remedies for infringement.

Effective Date

The Orphan Works Act of 2008 would have applied to infringements that commenced on or after January 1, 2009, except for infringing uses of pictorial, graphic, and sculptural works.[43] For the latter class of works, the limitations on liability would have applied to infringing uses that commenced on or after the earlier of: January 1, 2013, or the date on which the Copyright Office had certified at least two separate and independent, Internet-accessible electronic databases that allow for searches of copyrighted pictorial, graphic, and sculptural works.

Reports to Congress

H.R. 5889 directed the Register of Copyrights to report to Congress no later than December 12, 2014, on the implementation and effects of the limitation of liability for orphan works, including any recommendations for legislative change.[44] The Register was also directed to conduct an inquiry with respect to remedies for "small" copyright infringement claims, that is, those seeking limited amounts of money damages.[45] This requirement appeared to address the concerns of participants in the orphan works roundtables who expressed frustration at the expense of litigating a claim for copyright infringement. The Copyright Office was to invite public comment and conduct roundtables. At the conclusion, and not later than two years after the date of the enactment of the Orphan Works Act of 2008, the Office was to submit a report on this study to Congress, including such recommendations that the Register considered appropriate.

S. 2913, 110th Congress, Second Session, the Shawn Bentley Orphan Works Act of 2008

The Shawn Bentley Orphan Works Act of 2008 mirrored many of the provisions of H.R. 5889 although it lacked several of the House bill's features. S. 2913, as introduced, did not require an infringer who seeks to qualify for an orphan works limitation on remedies to file a Notice of Use with the Copyright Office (therefore, the bill also did away with the need for the Copyright Office to maintain an archive retaining such notices). S. 2913 also did not contain an exception for infringers of useful articles (meaning that such infringers would be able to qualify for the limitations on remedies), nor did it require a court to give consideration to the fact that a work is registered in determining reasonable compensation. S. 2913 did, however, explicitly extend the safe harbor exemption for paying reasonable compensation to museums in addition to those entities that the House bill had exempted (nonprofit educational institutions, libraries, archives, and public broadcasting entities).[46] S. 2913 also differed from H.R. 5889 in terms of the effective date of the limitations on remedies for infringers of pictorial, graphic, and sculptural works, offering a date of the earlier of: January 1, 2011, or the date on which the Copyright Office had certified at least two electronic databases that met the same criteria as the House bill.

Developments since the Introduction of H.R. 5889 and S. 2913

On May 6, 2008, the House Subcommittee on Courts, the Internet, and Intellectual Property approved by voice vote a manager's amendment in the nature of a substitute that included several substantive changes to H.R. 5889.[47] The amendment explicitly added museums to the list of entities that could qualify for the safe harbor of not having to pay reasonable compensation for infringing uses of orphan works, eliminating one difference with the Senate version of the orphan works legislation. Also, the manager's amendment instructed a court, before granting injunctive relief, to take into account any harm that the relief would cause the infringer due to the infringer's reliance on having performed a qualifying search for the copyright owner. Finally, the amendment clarified that the limitation on remedies for use of orphan works would not apply for an infringer who "fails to render payment, in a reasonably timely manner, of any reasonable compensation agreed upon by the owner of the infringed copyright *and the infringer*" (emphasis added to show the change made by the amendment). According to Representative Howard Berman who offered the amendment, this additional language helped to ensure that the copyright owner could not unilaterally "demand any amount of money, call it reasonable, and subject a user to damages."[48]

On May 15, 2008, the Senate Judiciary Committee approved by voice vote an amendment to S. 2913 in the nature of a substitute that was offered by its sponsors, Senators Leahy and Hatch.[49] The amendment added the "useful articles" exception that was provided in the House bill, thereby disallowing infringers who use useful articles from trying to claim the limitation on remedies. In addition, like the manager's amendment to the House bill, the amendment to S. 2913 instructed a court, before granting injunctive relief, to take into account any harm that the relief would cause the infringer due to the infringer's reliance on having performed a qualifying search for the copyright owner. The amendment also changed one of the potential effective dates with respect to the limitation on remedies for infringement of pictorial, graphic, and sculptural works, to January 1, 2013, matching that of the House version. The amendment also included a clarification similar to that in the House amendment, disallowing the limitation on remedies for an infringer who "fails to render payment of reasonable compensation in a reasonably timely manner *after reaching an agreement with the owner of the infringed copyright*" (emphasis added to show the change made to the original text of the bill). The substitute bill deleted a provision in the bill as introduced that would have denied museums, nonprofit educational institutions, libraries, archives, and public broadcasting entities the benefit of the safe harbor exemption from paying reasonable compensation if the owner of the infringed copyright proved (and the court found) that the infringer had earned proceeds directly attributable to the infringement. This provision still appeared, however, in the House bill.

On September 26, 2008, the Senate passed S. 2913 with an amendment[50] by unanimous consent.[51] The amendment made several modifications to S. 2913, including the following:

- Revised the "requirements for qualifying searches" section in a manner that directed the infringer to "undertake[] a diligent effort that is reasonable under the circumstances to locate the owner of the infringed copyright prior to, and at a time reasonably proximate to, the infringement."[52]
- Defined "diligent effort" to require, at a minimum, the following conduct: searching Copyright Office records and "reasonably available sources of copyright authorship and ownership information"; using technology tools, printed publications, and expert assistance; and searching databases including those available to the public through the Internet.
- Further defined "diligent effort" to encompass "any actions that are reasonable and appropriate under the facts relevant to the search, including ... facts uncovered during the search."[53]

However, the House took no action on S. 2913, and H.R. 5889 did not make it out of the House Judiciary Committee, before the end of the 110th Congress.

THE GOOGLE LIBRARY PROJECT AND ORPHAN WORKS

Although Congress has not yet passed orphan works legislation, recent developments in the private sector have further increased public interest in the orphan works problem. Specifically, questions have been raised about the fate of orphan works under the proposed settlement agreement that would resolve litigation regarding Google's proposal to scan, digitize, and index millions of print books in the collections of several major libraries, without seeking the permission of the copyright owners of those books.[54] Authors and publishers had brought a class action lawsuit to enjoin Google's reproduction of the books and to recover monetary damages for the company's alleged copyright infringement. A settlement agreement was first announced in October 2008, but the terms of the settlement prompted questions about Google's potential monopolization of book searching, the treatment of orphan works, protection of searchers' privacy, and the rights of foreign authors. A revised settlement agreement was submitted to the reviewing court in November 2009.

The agreement calls for Google to compensate rights holders for prior and future uses of their work. In addition, Google would be responsible for funding the creation and initial operations of a not-for-profit entity, called the Registry, which would represent rights holders in negotiating future uses of their content with Google.[55] However, some commentators are concerned that, with respect to books that may be orphan works, the absence of known or identifiable rights holders effectively means that these particular books digitized by Google could not be further used without the agreement of the Registry and/or Google.[56] Others have criticized the settlement for, in their opinion, "unilaterally giving all digital rights to orphan works to Google."[57]

Supporters of the settlement noted that many alleged orphan works are not, in fact, orphans, and that the settlement will prompt their owners to identify themselves, or reveal that the works are in the public domain, thus helping to resolve the orphan works problem.[58] Others argue that orphan works concerns are far less for books compared to other copyrightable subject matter:

> First, finding the rights owner of a book is not as daunting as many seem to believe. Books do not present the classic orphan works problem, photographs do. Photographs, both in the physical world and online, often become separated from their identifying information. This makes finding the rights owner a near impossibility. Books, however, always contain author and publisher information, and there's often a copyright registration record to help locate the rights owner. Second, although a copyright-protected book may have been published as long ago as 1923, the vast majority copyright-protected books in our libraries are far more recent.[59]

In a hearing before the House Judiciary Committee in September 2009, the Register of Copyrights offered her views about the proposed settlement's impact on future orphan works legislation:

> Congress should be particularly concerned about the settlement since it would interfere with the longstanding efforts of Congress and many other parties to address the issue of orphan works. The broad scope of the out-of-print provisions and the large class of copyright owners they would affect will dramatically impinge on the exclusive rights of authors,

> publishers, their heirs and successors. Such alteration should be undertaken by Congress if it is undertaken at all. Indeed, this Committee has already invested significant time in evaluating the orphan works problem and weighing possible solutions. That process is not over. The Google Book Settlement would frustrate the Committee's efforts and make it exceedingly difficult for Congress to move forward. A much more productive path would be for Google to engage with this Committee and with other stakeholders to discuss whether and to what degree a diligent search for the rights holder should be a precondition of a user receiving the benefits of orphan works legislation, or whether a solution that is more like a compulsory license may make sense for those engaged in mass scanning. Whatever the outcome, Congress is much better situated than the judiciary to consider such important and far- reaching changes to the copyright system.[60]

In defending the settlement, a representative of Google asserted that the settlement "is a strong complement to, and not a substitute for, orphan works legislation."[61] He disagreed with the Register's claim that the settlement would change copyright law:

> The settlement represents the resolution of a long and hard-fought litigation among multiple parties with divergent interests. The suggestion that the settlement usurps the role of Congress to set copyright policy because the suit took the form of a class action is flatly wrong. The settlement does not establish new copyright law; it is not even a determination on the merits of copyright law. All the settlement represents is the means by which the class of rightsholders decided to resolve the lawsuit.[62]

It remains to be seen whether the court's consideration of the settlement agreement in the Google Library Project litigation affects the course of potential orphan works legislation in the 111th Congress. The court granted preliminary approval of the revised agreement on November 19, 2009, and scheduled a final fairness hearing on it for February 18, 2010. The court will consider any objections from class members, as well as conduct an independent review of the proposed agreement, in determining whether to grant final approval.

ACKNOWLEDGMENTS

This chapter was originally prepared by Robin Jeweler, Legislative Attorney.

End Notes

[1] Copyright Office, *Orphan Works: Notice of Inquiry*, 70 FED. REG. 3739 (Jan. 26, 2005).

[2] The full report is available on the U.S. Copyright Office's website at http://www.copyright.gov/orphan/orphan-repo. pdf. Additional material, including the *Notice of Inquiry* and proceedings from the roundtable meetings, are also available there.

[3] *"Report on Orphan Works by the Copyright Office":* Hearing before the House Subcomm. on Courts, the Internet, and Intellectual Property, 109th Cong., 2d Sess. (2006); *Orphan Works: Proposals for a Legislative Solution:* Hearing before the Senate Comm. on the Judiciary, 109th Cong., 2d Sess. (2006); *Promoting the Use of Orphan Works: Balancing the Interests of Copyright Owners and Users:* Hearing before the House Subcomm. on Courts, the Internet, and Intellectual Property, 110th Cong., 2d Sess. (2008).

[4] U.S. CONST. ART. I, § 8, cl. 8. For more background information on copyright law, see CRS Report RS22801, *General Overview of U.S. Copyright Law*, by Brian T. Yeh.

[5] 17 U.S.C. § 106.

[6] In some cases, the Copyright Act prescribes terms for usage through compulsory licensing, but alternatives to traditional negotiated terms of usage are not discussed herein.

[7] *Notice of Inquiry,* 70 FED. REG. at 3741.

[8] *See* U.S. Copyright Office, *Circular 22: How to Investigate the Copyright Status of a Work* at http://www.copyright.gov/circs/circ22.pdf.

[9] A chart entitled *Copyright Term and the Public Domain in the United States,1 January 2010,* at http://copyright.cornell.edu/resources/publicdomain.cfm, provides an illustration of factors, such as publication, copyright notice, and renewal, that might apply in determining whether a work has entered the public domain.

[10] *See* U.S. Copyright Office, *Circular 15A: Duration of Copyright* at http://www.copyright.gov/circs/circ15a.pdf; *Circular 15t: Extension of Copyright Terms.* at http://www.copyright.gov/circs/circ15t.pdf.

[11] See, for example, online indices for a music performing rights organization (PRO) such as ASCAP at http://www.ascap.com or the Harry Fox Agency at http://www.harryfox.com/index.jsp, or photo clearing houses such as Photographers Index at http://www.photographersindex.com/.

[12] *See* U.S. Copyright Office, *Circular 15: Renewal of Copyright* at http://www.copyright.gov/circs/circ15.pdf.

[13] *See* H.Rept. 94-1476, 94th Cong., 2d Sess. 147 (1976) and S. Rept 94-473, 94th Cong., 2d Sess. 130 (1976). ("[The bill] takes a middle-ground approach in an effort to encourage use of a copyright notice without causing unfair and unjustifiable forfeitures on technical grounds.").

[14] 17 U.S.C. § 408.

[15] 17 U.S.C. § 411.

[16] 17 U.S.C. § 412.

[17] 17 U.S.C. § 502.

[18] 17 U.S.C. § 504.

[19] 17 U.S.C. § 505.

[20] Report on Orphan Works (hereinafter Report) at 2. "These include situations where the user contacted the owner, but did not receive permission to use the work, either because the owner did not respond to the request, refused the request, or required a license fee that the user felt was too high."

[21] Specifically, the "idea/expression" dichotomy, 17 U.S.C. § 102(b), prohibits copyright protection for any idea, procedure, process, system, method of operation, concept, principle, or discovery.

[22] Report at 53, citing at note 123, Eldred v. Ashcroft, 537 U.S. 186, 219-21 (2003).

[23] *Id.* at 36-40.

[24] "In considering legislative solutions to the orphan works problem it is important to keep in mind the requirements of the international instruments to which the United States has agreed: exercise and enjoyment of a copyright right cannot be conditioned on a formality, any exceptions or limitations on copyright must conform to the three-step test [under international treaty obligations], and the effect on the owner's remedies must comply with the various remedy rules." *Id.* at 68.

[25] *Id.* at 69.

[26] *Id.* at 127.

[27] 17 U.S.C. § 107.

[28] Roy de Forest, *County Dog Gentleman,* 1972, San Francisco Museum of Modern Art.

[29] *See* statement of David P. Trust, CEO of Professional Photographers of America before the House Judiciary Committee (March 8, 2006), *supra* note 4; statement of Victor Perlman, General Counsel of the American Society of Media Photographers, and statement of Brad Holland, Founding Board Member, Illustrators' Partnership of America, before the Senate Judiciary Committee (April 6, 2006), *supra* note 4.

[30] Sec. 2(a) of H.R. 5889, adding a new § 514(b)(1)(A).

[31] Sec. 2(a) of H.R. 5889, adding a new § 514(b)(2)(A).

[32] Sec. 3(a) of H.R. 5889.

[33] Sec. 2(a) of H.R. 5889, adding a new § 514(c)(1)(A).

[34] Sec. 2(a) of H.R. 5889, adding a new § 514(a)(4).

[35] Sec. 2(a) of H.R. 5889, adding a new § 514(c)(1)(C).

[36] However, H.R. 5889 allowed the owner of the infringed copyright to try to recover any proceeds directly attributable to the infringement if the copyright owner can prove, and the court finds, that the infringer has earned such proceeds. Sec. 2(a) of H.R. 5889, adding a new § 514(c)(1)(B).

[37] Such notice must be written, and include at a minimum the following information: the name, address, and telephone number of the owner of the infringed copyright; the title of the infringed work or a detailed description of it; and information from which a reasonable person could determine the validity of the copyright owner's claim of ownership and alleged infringement. Sec. 2(a) of H.R. 5889, adding a new § 514(a)(2).

[38] Sec. 2(a) of H.R. 5889, adding a new § 514(b)(1)(B).

[39] Sec. 2(a) of H.R. 5889, adding a new § 514(b)(4).

[40] Sec. 2(a) of H.R. 5889, adding a new § 514(c)(2).

[41] Sec. 2(a) of H.R. 5889, adding a new § 514(d).

[42] 17 U.S.C. § 101 (definition of "useful article").

[43] Sec. 4(a) of H.R. 5889.
[44] Sec. 5 of H.R. 5889.
[45] Sec. 6 of H.R. 5889.
[46] Sec. 2(a) of S. 2913, as introduced, adding a new § 514(c)(1)(B).
[47] Carey Lening, *Amended Orphan Works Reform Bill Advances Through IP Subcommittee,* Patent, Trademark, & Copyright J., May 9, 2008, at 47.
[48] Andrew Noyes, *'Orphan Works' Measure Moves to Full Committee,* CongressDailyAM, May 8, 2008.
[49] Andrew Noyes, *Senate Panel Approves 'Orphan Works' Copyright Bill,* CongressDailyPM, May 15, 2008.
[50] S.Amdt. 5669, proposed by Senator Whitehouse for Senator Kyl.
[51] *Orphan Works Legislation Passes Senate By Unanimous Consent, Moves to House,* Patent, Trademark, & Copyright J., Oct. 3, 2008, at 754.
[52] S. 2913, as engrossed, § 2(a) (adding new 17 U.S.C. § 514(b)(2)(A)(i-v)).
[53] S. 2913, as engrossed, § 2(a) (adding new 17 U.S.C. § 514(b)(2)(A)(ii)).
[54] For a comprehensive overview of this litigation and the proposed settlement, *see* CRS Report R40194, *The Google Library Project: Is Digitization for Purposes of Online Indexing Fair Use Under Copyright Law?*, by Kate M. Manuel.
[55] Authors Guild, Inc. v. Google Inc., Settlement Agreement, Case No. 05 CV 8136-JES (S.D.N.Y. Oct. 28, 2008).
[56] *See* Sherwin Siy, *The New Google Book Settlement: First Impressions on Orphan Works,* Nov. 17, 2009, *at* http://www.publicknowledge.org/node/2770 ("Google should not be the sole entity able to license the display of orphan and unclaimed works. Nothing in the new settlement agreement seems to change that dynamic. Google and the plaintiffs have previously stated that they do not believe that the Book Rights Registry—a settlement-created entity that is supposed to represent the interests of authors—would be able to grant other entities the same ability to legally display orphan and unclaimed works...").
[57] Letter from the National Writers Union, the American Society of Journalists and Authors, and the Science Fiction and Fantasy Writers of America, to Congressional Authors, Jan. 6, 2010, *available at* http://www.openbookalliance.org/wp-content/uploads/2010/01/Writers-to-Congressional-Authors-Letter.pdf.
[58] *See, e.g.,* Nathan Pollard, *Social Justice IP Panel Says Google Book Settlement Will "Level the Playing Field,"* 78 Patent, Trademark & Copyright J. 408 (July 31, 2009).
[59] *"Competition and Commerce in Digital Books":* Hearing before the House Judiciary Comm., 111th Cong., 1st Sess. (2009) (statement of Paul Aiken, Executive Director, the Authors Guild, at 8.).
[60] *"Competition and Commerce in Digital Books":* Hearing before the House Judiciary Comm., 111th Cong., 1st Sess. (2009) (statement of MaryBeth Peters, the Register of Copyrights, at 7-8).
[61] *"Competition and Commerce in Digital Books":* Hearing before the House Judiciary Comm., 111th Cong., 1st Sess. (2009) (statement of David Drummond, Senior Vice President of Corporate Development and Chief Legal Officer, Google, Inc., at 6).
[62] *Id.* at 7-8.

In: Digital Books: Competition and Commerce
Editor: Oliver A. Hagen ISBN: 978-1-61728-299-7

Chapter 3

STATEMENT OF PAUL AIKEN ON THE GOOGLE BOOK SETTLEMENT

Paul Aiken

I. EXECUTIVE SUMMARY

Overview

My name is Paul Aiken. I'm the executive director of the Authors Guild, the largest society of published authors in the U.S., representing more than 8,500 book authors and freelance writers. Our members represent the broad sweep of American authorship, including literary and genre fiction, nonfiction, trade, academic, and children's book authors, textbook authors, freelance journalists and poets.[1] Guild members have won countless honors and all major literary awards, including the Nobel Prize for Literature.[2]

The Authors Guild promotes the professional interests of authors: we're advocates for effective copyright protection, fair contracts, and free expression.

The Challenges Facing Print Media

It's a pleasure to be here before this committee, at this moment in book publishing history. Never in the Authors Guild's long history has its straightforward mission – to maintain writing as a viable livelihood – been so daunting. The digital environment is brutal for print media. As we meet here today, the newspaper industry is dying. Credible estimates say that one newspaper is closing each week in America. The magazine industry isn't much better off, as week by week we see venerable publications shrink in size and ambition. The loss to our society from the collapse of these industries is immeasurable.

The book industry, happily, has to date fared better than our colleagues in the print media. This is, no doubt, partly due to our medium: print books are still superior in almost every way to their electronic counterparts. That advantage is rapidly fading, however; our transition to digital form is underway, and things change quickly in a digital environment.

The portents are not encouraging. Finding a sustainable business model for creative work in digital form seems nearly impossible on the Internet: if piracy doesn't get you, the aggregators will.

We'll likely need many things to go right to avoid the fate of our colleagues in the print industries.

The Opportunity

Yet, there are reasons for optimism. One of those reasons is our settlement with Google, which brings us here today. That settlement promises to address one of the oldest and most vexing of market failures: the loss to the commercial market of out-of-print books.

If you had asked knowledgeable people a couple of years ago whether we were close to delivering a near universal library to public libraries, colleges, and universities across the country, they would have scoffed. The technical challenges seemed too daunting, the rights clearance issues insurmountable, the passions stirred by the ongoing copyright wars far too intense for such a result to be achieved in the foreseeable future.

But here we are, on the cusp of that extraordinary achievement: the marriage of much of our collective library with the Internet. The benefits to readers, students, and scholars would be profound. Here are a few:

1. The settlement would turn every library into a world-class research facility, by offering every public library building in the U.S. – all 16,500 of them – a free portal to millions of out-of-print books. The settlement would also offer a free portal to that same vast database of out-of-print books to more than 4,000 higher education institutions, from community colleges to our most elite universities.
2. Students and professors at colleges with the most modest of endowments would find an important part of the academic playing field had come to level, as they gain full access from every computer on campus to a library exceeding that of the finest Ivy League schools.
3. The visually and reading impaired would find the stacks of libraries open up to them as never before, gaining access to orders of magnitude more books than they currently have.
4. The settlement would offer anyone online in the U.S. free "preview" access to hundreds of millions of pages of text (up to 20% of each book).[3] Readers from their own home computers would be able to review hundreds of accounts of the Battle of Vicksburg, or of the beginnings of the Industrial Revolution, or of the sources and interpretation of Moby Dick, at no charge. Should a reader find one book particularly compelling, she could buy access to the entire book. Access to public domain books is free, of course, and authors controlling the rights to their books can choose to give away access for free.

Authors and publishers are willing to make this deal for several reasons. We of course hope to profit from the market that's created. We would like to have the Internet work for us, creating a market of the previously unmarketable. We also have a vital interest in keeping books central to our students, scholars, and culture. We're confident that making this vast library available online will help do just that.

Authors have another strong interest in making this deal work: authors need libraries. Libraries fuel their work. Authors of every type read, reinterpret and rely on their fellow authors, and those who have come before them. This is true of the scholarly writer and of the author of popular nonfiction. It's as true for authors of books for children as it is for authors of books for adults. Authors of literary fiction also rely heavily on those who've come before them. The creative expressions are new, but many of the ideas underlying literary works are eternal. Writers of genre fiction are no exception. Romance writers read romance novels and other works and offer their own interpretation and variations on the romance theme.

Authors, in short, want not only to realize the untapped value of their out-of-print works, they want access to this new, vast online library so that they can more easily create new works that readers will value.

The Opposition

Opposition to the settlement falls into several broad categories. We'll address two important objections in this Executive Summary.

Objection #1: Copyright doesn't permit a system that asks authors and publishers to specifically exclude their out-of-print works from uses negotiated on their behalf.

This simply isn't so. There is ample precedent around the world for dealing with market failures in copyright in precisely this way. For example, Germany today operates a system that nearly parallels the one the settlement would put in place. The German system allows for

(a) routine copying of out-of-print works written by foreign and domestic authors,
(b) including routine copying of "orphan" (unclaimed) works, and
(c) without regard to whether those authors and publishers have expressly approved those uses.

The German system, however, actually denies authors and publishers any ability to exclude their works.

The German system goes even further, allowing the copying of in-print works without the permission of the author or publisher (in our settlement, the author and publisher must both approve of any displays of in-print works). Germany's photocopy licensing system is perfectly legal, and meets with the norms of international copyright law.

In fact, this is the typical way for countries to deal with the market failure represented by the unlicensed photocopying of copyrighted materials, although many countries allow authors and publishers to exclude their works from such licensing. Other countries with similar photocopy licensing systems include Australia, Canada, the United Kingdom and the Nordic countries, among many others. The market for photocopy licensing often fails without intervention, because the transaction costs of the license, including the labor costs of the licensee, are simply too high relative to the value of the individual copy.

The inability to license out of print works to colleges, libraries and individual users presents a market failure on an epic scale. Here, as with photocopy licensing, a major component of the transaction costs involve rights clearance issues. Another impediment is the sheer scope of the project, and the capital and technological resources it demands. This settlement, with a financially strong and sophisticated technology partner, addresses the market failure. The societal value in bringing these works back to the market is incalculable, but until our settlement there was no practical way to do so.[4]

Objection #2. The settlement inappropriately permits the use of unclaimed (orphan) works.

No issue has been more misunderstood or misreported regarding this settlement than the unclaimed or "orphan" works issue.

The primary misconception is the size of the problem: it's much smaller than has commonly been reported, for several reasons. First, finding the rights owner of a book is not as daunting as many seem to believe. Books do not present the classic orphan works problem, photographs do. Photographs, both in the physical world and online, often become separated from their identifying information. This makes finding the rights owner a near impossibility. Books, however, always contain author and publisher information, and there's often a copyright registration record to help locate the rights owner. Second, although a copyright-protected book may have been published as long ago as 1923, the vast majority copyright-protected books in our libraries are far more recent.[5]

Another major misconception is the failure to recognize that countries around the globe are already dealing with the orphan works issue in a productive way. The photocopy-licensing systems in other English-speaking countries permit the use of orphan works. This, as previously discussed, is a natural result of those nations' attempts to cope with the market failure represented by unlicensed photocopying of copyrighted works. As the licensing societies collect photocopy royalties and start to cut checks to authors, word spreads, quickly, and authors step forward to register themselves. The pool of unlicensed works shrinks. The licensing societies are duty-bound to actively seek out authors[6] for whom they have money. Year by year, they locate more and more authors, and the orphan works problem diminishes further.

We have some experience with this, since we helped found and long provided financial support to the Authors Registry, an independent, non-profit, rights-payment agency. The Authors Registry collects photocopy and other use fees from overseas, particularly from photocopy uses in the U.K., and pays authors in the U.S. the amounts due them. A sample of our success in paying authors of out-of-print works last year suggests that we reach 85% of such authors. The success rate of larger, more developed systems – such as that of the Authors' Licensing and Collecting Society (ALCS) in the U.K. – demonstrates that even higher success rates are possible. (The ALCS, representing more than 30,000 published writers, is an enthusiastic supporter of the settlement, for good reason. They know it can work, because they've achieved great success finding and paying photocopy revenues to authors of out of print books.)

Thus, this settlement presents a practical solution for the problem of orphan works for books. When an author is identified, then requests for all kinds of other uses – for permission to use an excerpt from the author's work, reprint it, or to translate it into a foreign language – can be relayed to the author or the author's agent and acted upon.

The orphan works issue is far smaller, and far more tractable, than some objectors would have you believe. The settlement itself is a big part of the solution.

Conclusion

We urge this committee to recognize this settlement for what it is: the outstanding result of a rare and productive truce in the copyright wars, negotiated by strong-willed and pragmatic representatives of the author, publisher, and library communities and a sophisticated technology partner.

To a dispassionate observer, we believe the solution presented by this settlement is how a rational, useful market for out-of-print books should operate in the digital age. The means of getting there, a class-action settlement, may be novel, but that shouldn't distract us from the great good – for readers, students, scholars, authors and publishers – that this settlement accomplishes. Similar systems, inevitably, will develop around the world.

This settlement doesn't pre-empt congressional action, but there's no need to act now, before we see how well this solution works in the real world. We suspect many of the concerns – including all of the major objections – will prove unwarranted as this settlement goes into operation. There's no need to fix that which likely isn't broken at all.

Allowing this opportunity to slip through our grasp would be a tragic loss to all those who value the riches stored in our nation's libraries.

II. Introduction to the Settlement

In 2004, Google Inc. ("Google") announced that, as part of its Google Library Project (or "GLP"), it would reproduce millions of copyrighted books located in U.S. libraries and display "snippets" of those books on its website. These books had been published within and outside the United States and their copyright owners included publishers and authors domiciled in both the United States and around the world. In copying and displaying content from books as part of the GLP, Google did not seek the permission of the copyright owners of those books.

The Settlement resolves two copyright infringement actions brought against Google by authors and publishers that alleged that Google's unauthorized copying of books and display of snippets without permission constituted copyright infringement under the U.S. copyright laws. Google defended its actions as non-infringing fair uses, as authorized by 17 U.S.C. § 107.

At the end of October 2008, after a negotiation that lasted more than two years, the author and publisher plaintiffs and Google announced that they had reached a comprehensive settlement of the litigation. The Settlement avoids further discovery and litigation, which could have been protracted.

The Settlement was carefully negotiated to protect the copyright interests of, and to provide meaningful benefits to a class of authors (and their heirs) and publishers of books covered by the Settlement (essentially, books published on or before January 5, 2009). One of the Settlement's crowning achievements for the class -- and for the United States public -- is that it breathes new life into millions of out-of-print books, which until now have been relegated to the dusty stacks of university libraries. The Settlement also provides new marketing and revenue opportunities for rightsholders of in-print books (should they choose to take advantage of them), in a way that does not harm the existing markets for those books.

If approved, the Settlement will provide the following material benefits to authors and publishers:

- Payment to Rightsholders: Google will be authorized to use class members' works in several revenue models, and will pay 63% of the revenues earned from its exploitation to a Book Rights Registry (the "Registry"), for disbursement (after an administrative charge) to the Rightsholders of those works;
- Rightsholders Retain Control: The Settlement is non-exclusive and does not involve any transfer of copyright ownership interests or any other property interests to Google. Additionally, Rightsholders at all times will retain control over their works, with the ability to determine the extent to which their works are to be included in or excluded from Google's uses, and to license their works to others;
- Compensation to Rightsholders for Past Alleged Infringement: Google will pay at least $45 million to compensate class members whose works Google digitized without permission before May 6, 2009.
- Establishment of Registry: Google will pay $34.5 million both for notice and settlement administration costs and to fund the Registry, a non-profit organization that will be managed by a Board representing authors and publishers, that will locate rightsholders, maintain a database of their contact information, collect and pay revenues to the class for Google's use of copyrighted works through this Settlement, and otherwise protect and represent the interests of the class.

As explained in more detail below, the Settlement provides extraordinary and previously unattainable benefits to authors and publishers in the United States and all over the world, along with remarkable benefits to the reading public, students, scholars and researchers in the United States.

III. The Material Terms of the Settlement

A. The Settlement Class and Sub-Classes.[7]

The Settlement Class is defined as follows:

All persons or entities that, as of January 5, 2009, have a Copyright Interest in one or more Books or Inserts. All Settlement Class members are either members of the Author Sub-Class or the Publisher Sub-Class, or both. Settlement Agreement ("SA") § 1.142.

- **Author Sub-Class**
  All members of the Settlement Class who are authors, and their heirs, successors, and assigns, and any other members of the Settlement Class who are not members of the Publisher Sub-Class. SA § 1.14.

- **Publisher Sub-Class**
  All members of the Settlement Class that are (a) companies that publish books, and their exclusive licensees, successors, and assignees, and (b) companies that publish periodicals and have a Copyright Interest in one or more Inserts, and their exclusive licensees, successors, and assignees. SA § 1.120.[8]

B. The Parties

The Representative Plaintiffs for the Author Sub-Class are authors Paul Dickson, Joseph Goulden, Daniel Hoffman, Betty Miles, and Herbert Mitgang. SA § 1.125. The Representative Plaintiffs for the Publisher Sub-Class are The McGraw-Hill Companies, Inc., Pearson Education, Inc., Penguin Group (USA) Inc., Simon & Schuster, Inc., and John Wiley & Sons, Inc. *Id.* The Authors Guild and the AAP are "Associational Plaintiffs." SA § 1.12. The Associational Plaintiffs participated in the litigation to advance the rights and interests of the Author Sub-Class and Publisher Sub-Class, respectively, and served as consultants to counsel for the respective sub-classes.

Google was the sole defendant in the cases. In addition, though not parties to the litigation, the libraries of the University of Michigan, Stanford University, and the University of California took part in the settlement negotiations. They, along with a large number of other libraries, have entered into digitization agreements with Google, to allow their collections to be digitized. Numerous university libraries plan to participate in the GLP as well. A list of libraries that are authorized to provide books to Google for copying, and to receive digital copies, can be found at Attachment G of the SA.

C. Benefits to Authors and Publishers

1. Revenues from Google's Use of Books and Inserts

The Settlement authorizes Google, on a non-exclusive basis, to digitize Books and Inserts, to develop a searchable electronic books database, and to display and make commercial use of Books and Inserts, including the following "Display Uses":

- sell institutional subscriptions to schools, corporations, and government offices ("Institutional Subscriptions") (SA § 4.1);
- sell online access to Books to consumers ("Consumer Purchases") (SA § 4.2);
- display "previews" (up to 20% of a Book (SA § 4.3)) and "snippets" (several lines of a Book (SA § 1.147)) in order to spur Book sales and earn advertising revenues;
- place advertisements next to Book text (SA § 4.4); and

- provide free Public Access Services to all public and school libraries, with any printing from these services subject to a per-page fee (SA § 4.8).[9]

Note that all authorizations, including the right to digitize books and to display and make commercial uses of Books and Inserts, are confined to the United States and Google obtains no rights under the Settlement under any law other than United States law.

Revenues earned from Display Uses will be split 63/37 in favor of the Rightsholders. SA § 4.5. ("Rightsholders" are members of the Settlement Class that do not opt out of the Settlement.) Google will pay 63% of these revenues to the Registry, for disbursements to the Rightsholders.

2. Classification of Books

The Settlement differentiates between Commercially Available (generally, in- print) and not Commercially Available (generally, out of print) Books. The purpose of this distinction is to ensure that Google's Display Uses do not cannibalize the existing markets for Commercially Available books. Thus, the Settlement treats Commercially Available Books differently from Books that are not Commercially Available (i.e., Books that are not being marketed or sold by the publishers of the Books). SA §§ 3.2-3.3. Specifically, Google may not make any Display Uses of Commercially Available Books – both the revenue models described above and "snippet display" – unless both the author and publisher authorize Google to do so. *Id.* In contrast, Books that are not classified as Commercially Available will, by default, be included automatically in all Display Uses. At any time, however, the Rightsholder of a Book can change the default and instruct Google to exclude any Book from any one or more Display Uses. *Id.*

Google will initially determine whether or not a Book is Commercially Available as of January 5, 2009 (the Notice Commencement Date). SA § 3.2(d) (referred to in the SA as the "Commercially Available" classification). Google and the Registry will continue to be assessing whether Books are Commercially Available using multiple sources of information. Rightsholders who claim their Books will be informed if a change is made in the Commercially Available classification of those Books.

Rightsholders can challenge the determination of whether a Book is Commercially Available, and any disputes with Google will be resolved through the Settlement's arbitration procedures (*see* SA Article IX). SA § 3.2(d)(iv).

3. Rightsholders' Do-Not-Digitize/Removal Rights

Rightsholders can tell Google not to digitize their Books, or if already digitized, to remove their Books from Google's database and from the digital copies provided by Google to participating libraries. SA §§ 1.124, 3.5(a)(i). Google and the libraries must honor all do-not-digitize/removal requests made by April 5, 2011; thereafter requests will be honored only if the Books have not already been digitized. SA § 3.5(a)(iii).[10]

4. Rightsholders' Exclusion Rights

Even if they do not remove a Book from Google's and the libraries' databases, Rightsholders at any time can exclude their Books (or portions thereof) from any or all Display Uses. SA § 3.5(b). Insert Rightsholders at any time can exclude Inserts (or portions thereof) from all (but not less than all) Display Uses. *Id.* Rightsholders can change their mind at any time and turn Display Uses back on for some or all of their Books and Inserts. *Id.*

5. Rightsholders' Pricing Rights for Consumer Purchase

Rightsholders have two options under the Settlement for setting the sale price of their Books made available in the Consumer Purchase revenue model: they can set the price themselves (Specified Pricing) (SA § 4.2(b)(i)(1)) or they can rely on the "Settlement-Controlled Price" developed by Google. (SA § 4.2(b)(i)(2)). Settlement- Controlled Prices will be algorithmically designed by Google to find the optimal revenue- maximizing price for each Book. (SA §§ 4.2(b)(i)(2) & (c)).

Plaintiffs have heard from a number of academic and other authors (e.g., university professors) who are interested in making their books freely available to others through the Settlement. The Settlement Agreement enables Rightsholders to do so. The Settlement Agreement will support Creative Commons licenses. In addition, Rightsholders can set the sale price for their Books at zero. In those situations neither Google nor the Registry will receive revenues from any Consumer Purchase of those Books.

6. Establishment of the Book Rights Registry, and Payment for Notice and Claims Administration

Google has agreed to pay $34.5 million to fund the newly formed Book Rights Registry, to pay for notice to the Settlement Class and to pay for claims administration services. SA § 2.1(c). The Registry will be an independent not-for-profit entity, and all funds received by the Registry will be for the benefit of Rightsholders. SA § 6.2(a). The Registry will have a Board of Directors composed at all times of an equal number of Author Sub-Class and Publisher Sub-Class representatives. SA § 6.2(b). Thus, the Registry will not, as some have suggested, by controlled by Google. The Registry will establish and maintain a database of Rightsholders' contact information and information regarding Rightsholders' Books and Inserts. SA § 6.1(b). The Registry will locate Rightsholders, identify and coordinate payments to Rightsholders, and otherwise represent the interests of Rightsholders under the Settlement Agreement. *Id.* The Registry will able, with express rightsholder approval, to do deals with anyone, including competitors of Google.

The Registry will be funded at first by Google's $34.5 million payment (net of notice and claims administration expenses). SA §§ 1.6, 2.1(c), 5.2. Thereafter, the Registry will also be funded by an administrative fee taken as a percentage of Rightsholders' revenues. SA, Attachment C (Plan of Allocation) § 4.2.

7. Payment for Already Digitized Books

To settle plaintiffs' claims for actual infringement, Google has agreed to pay Rightsholders a minimum of $45 million for Books and Inserts digitized without permission as of May 5, 2009 (i.e., the original opt-out deadline). SA §§ 2.1(b), 5.1. For every such work that is claimed by January 5, 2010, Google will make a "Cash Payment" of at least $60 per Principal Work, $15 per Entire Insert, and $5 per Partial Insert.[11]

If more than $45 million is required to pay all of the eligible claims for Cash Payments, then Google will pay the additional funds necessary to make all such Cash Payments, with no cap on such additional payments. SA § 5.1(b). If the total amount of all eligible claims is less than $45 million, the claiming Rightsholders can receive up to $300 per Principal Work, $75 per Entire Insert, and $25 per Partial Insert. Any remaining funds thereafter from the $45 million will go to the Registry for operations. SA, Attachment C (Plan of Allocation) § 3.2.

D. Author-Publisher Procedures

Because, for most Books, authors and their publishers each have copyright interests, the authors' and publishers' respective interests are addressed in the Author- Publisher Procedures (SA, Attachment A). Among other things, the Author-Publisher Procedures set forth the authors' and the publishers' respective rights concerning in-print and out-of-print Books under the Settlement.

1. In-Print Books

With respect to in-print books (other than works-for-hire), both the author and the publisher must agree to include the Book in the GLP; if they do not agree, it will not be included. Author-Publisher Procedures ("A-P") § 5.1. Cash Payments for in-print Books, as well as all revenues earned from Google's future uses of in-print Books, will be paid to the publishers of the Books, who will then pay the authors under the book publishing contract between the author and the publisher. A-P § 5.5. If an author wishes to challenge the revenue split offered by the publisher under that contract, the dispute may (except for Educational Books) be resolved in the Settlement's arbitration process. *Id.* The A-P also sets forth the respective rights of authors and publishers as to exclusion, removal, pricing, and changes in Display Uses of in-print Books. A-P §§ 5.2-5.4.

2. Out-Of-Print Books

The A-P includes provisions concerning exclusion, removal, control of pricing, and changes in Display Uses of out-of-print Books, which vary depending on whether the Book is (1) a work-for-hire under United States copyright law (100% of Cash Payments and revenues paid by the Registry directly to the publisher), or (2) reverted (100% of Cash Payments and revenues paid by the Registry directly to the author). A-P § 6.1.

Cash Payments and future revenues earned for unreverted out-of-print Books will be split between the author and publisher of the Book as follows: (a) for Books first published prior to 1987, the Registry will pay 65% to the author and 35% to the publisher; and (b) for Books first published during or after 1987, the Registry will pay each of the author and the publisher 50%. A-P § 6.2(c).[12] In addition, for Books that are not reverted and are not works-for-hire, in general, both the author and the publisher have the right to manage the Books. (*See* A-P for further details concerning the A-P.)

E. Other Provisions

1. Public Access Service

The Parties have agreed to provide, upon request, free access to the entire subscription database at a computer terminal in every public library building in the United States that requests one and at least one computer terminal at each not-for-profit higher educational institution. SA §§ 1.66, 1.119, 4.8. Anyone in any urban or small town library building in the U.S. could have free, full access to the entire database of Books.

2. Non-Display Uses

In addition to the Display Uses, Google will be permitted to make "Non-Display Uses" of Books, including full-text indexing (without displaying the text), geographic indexing, algorithmic listings of key terms for chapters of Books, and other internal research. SA §§ 1.91, 3.4.

3. Security Provisions

Google and plaintiffs developed security standards to prevent security breaches and unauthorized use of Google's books database. SA §§ 8.1-8.2 and Attachment D. The SA provides remedies to Rightsholders for security breaches and unauthorized access. SA §§ 8.3-8.7.

4. Dispute Resolution Mechanism

If disputes concerning the Settlement arise among Rightsholders and Google or the libraries, they will be subject to arbitration. Examples are disputes (1) between authors and publishers of the same Books; (2) over claimed security breaches; (3) over whether a Book is in-print, out-of-print, or in the public domain; and (4) over whether Google or a participating library has made a use of a Book that is not authorized under the Settlement. *See generally* SA Article IX. The Settlement Agreement also provides that the court will have continuing jurisdiction over other disputes concerning the parties' obligations under the Settlement Agreement. SA § 17.23. Disputes between publishers are not subject to arbitration.

5. Non-Exclusive Rights

The authorizations granted to Google under the Settlement Agreement are non-exclusive only. Rightsholders retain all their rights to use and license their Books in any way, including ways identical to those authorized to Google. No authorization under the Settlement constitutes a transfer of any copyright ownership interest in any Book or Insert. SA §§ 2.4, 3.1(a).

IV. COPYRIGHT ISSUES

The Settlement has been the subject of substantial discussion. I would like to elaborate on my remarks in the Executive Summary regarding the Settlement's treatment of so-called "orphan works," and I will then address the potential effect of including foreign rightsholders in the Settlement Class on U.S. foreign relations.

A. The Question of "Orphan Works"

A number of critics of the Google Book Search Settlement have complained that the Settlement will confer on Google a monopoly over "orphan works." These critics have stated – without any evidence – that "orphan works" will include millions of books, comprising anywhere from 50% to 70% of the books covered by the Settlement. These percentages are over-inflated, principally because at least some of the critics appear to have equated the term "out-of-print books" with "orphan works," which is erroneous, and, further, because they incorrectly assume that the rightsholders of out-of-print books are either unknown or cannot be found.

The term "orphan works," however, has no meaning under the Copyright Act; the Act only recognizes works only as in-copyright or not in-copyright. That term, however, has been typically understood and used – both in the context of legislative proposals considered by the Congress, including this Committee, over the last several years, as well as by the Copyright Office – to refer to a work for which the copyright owner cannot be identified or found (such as a photograph with no attributed photographer).

Contrary to the use of that term by the Settlement's critics, we start with the principle that published books found in U.S. libraries – unlike many other types of copyrighted works – are generally quite unlikely to be orphan works. They are published works. They have identifiable authors and publishers. Published books include readily accessible information as to their author, publisher and date of publication.

Moreover, as to all United States works, the Settlement only covers those that are "registered" with the United States Copyright Office. So, at least with respect to United States works, the Copyright Office registration is available as an important starting point in attempting to identify and locate the current copyright owner. There is no reason to believe that a significant number of the copyright owners of books covered by the Settlement cannot be identified and found – by anyone, including would-be competitors of Google – if due diligence efforts are made to do so.

Prior to the Settlement, at least for purposes of large scale commercial exploitation, few such efforts have been made. That is because there has been no commercial market for a collection of out-of-print books, and thus no incentives to identify rightsholders of out-of-print books or for them to come forward to associate themselves with their works for purposes of licensing others to use them. Now, for the first time, the settlement provides just such a product, and brand new incentives for rightsholders of out-of-print books to participate. Not only have meaningful, ambitious efforts been made to locate these rightsholders, those efforts will be continued into the future by the Registry. In addition and importantly, the prospect of earning money under the Settlement – and the very availability of revenues generated by the Settlement – will provide powerful incentives for rightsholders to come forward to claim their works. Once they do so, of course, such works would have identifiable copyright owners and could not be considered to be "orphans."

First, approval of the Settlement required the parties to undertake an unprecedented, worldwide Notice Program that was designed to reach as many members of the Settlement Class as possible. As a result of the Notice Program, many rightsholders of out-of-print books have already claimed their books. Because there is no deadline to claim one's books through the Settlement, it is reasonable to expect that the numbers of claimed out-of-print books will grow exponentially.

Second, one of the Registry's core missions will be to locate rightsholders of outof-print books that have not yet been claimed. This is for the purpose of having the Registry assist them in claiming their works and, ultimately, to pay to them the revenues they are owed under the Settlement's revenue models.

Also, as noted above, the Registry will want to enter into licensing arrangements with others, not just Google. For this purpose, the Registry and claiming rightsholders will want as many rightsholders as possible to come forward and authorize the Registry to include their books in those arrangements. In this way, there will be a more robust set of books available for license, which will redound to the benefit of all rightsholders..

The goals of the Notice Program and the Registry are achievable precisely because the settlement creates meaningful incentives for copyright owners of out-of-print works to claim their books. The new services authorized by the Registry (which include the subscriptions and consumer purchase options offered by Google, as well as alternative, even competing products that others may establish) will begin earning rightsholders new revenue.

Most authors write for two reasons, for their books to be read, and to be compensated. Because the Settlement has given new exposure and commercial life to out-of-print books, authors are more likely to claim their books and the Registry will have a greater chance of finding them.

What does that mean for the so-called "orphan books?" We've already learned through the Settlement's claiming process that many out-of-print books' "parents" are alive and claiming their books. And, as "parent" rightsholders claim their books, the number of books that might arguably be considered "orphans" will be dramatically reduced.

Significantly, the Registry will maintain a publicly accessible database of which books are claimed. The database will also make public who has claimed those books (except where the claimant has asked that his or her name not be disclosed). This database will make it far, far easier than at present for anyone to identify rightsholders of books and obtain permission to use them. The experiences of the Authors Registry and the ALCS, reported in the Executive Summary, bear this out. Thus, where it might once have been difficult to find the copyright owners of books it will now become much more feasible to license out-of-print books, either through the Registry or directly from their copyright owners.

Of course, there may still be books whose rightsholders prove difficult to find. The Settlement is not a panacea. But the facts suggest, contrary to the gloomy scenarios painted by the critics, that the number of such books ultimately will be quite low.

B. The Settlement is Wholly Consistent with the Internationa Obligations of the United States

Questions also have been asked as to whether the Settlement complies with the international treaty obligations of the United States. These issues have been raised under two provisions of the 1971 Berne Convention for the Protection of Literary & Artistic Works (the "Berne Convention"), to which the United States adhered effective March 1, 1989.[13] The first provision is that of national treatment, found in Article 5(1) of the Berne Convention. The second provision is Article 5(2) of the Berne Convention, which bans the imposition of certain "formalities." Examining both provisions, we believe that the Settlement is fully consistent with our country's treaty obligations.

To begin, the Berne Convention is not itself self-executing in the United States. In the Berne Convention Implementation Act of 1988 (BCIA), which implemented the Berne Convention,[14] Congress was crystal clear that the Berne Convention itself does not create any rights or obligations under U.S. law.[15] Thus, if any provision of the enacted laws of the United States is inconsistent with our country's Berne Convention obligations, the only remedy is for another *country* to take the United States to the International Court of Justice, or to invoke the dispute resolution provisions set forth in the GATT, leading possibly to a WTO panel. With respect to the pending approval of the Settlement, the District Court must follow the Copyright Act, and it has no authority to deviate from it in an effort to comply with its understanding of the United States' treaty obligations.

1. National Treatment

The Settlement is fully consistent with the national treatment principle of Article 5(1) of the Berne Convention,[16] which Congress implemented in the BCIA, and is now enshrined in Section 104 of the Copyright Act, as amended. In the United States, that principle "simply assures that if the law of the country of infringement applies to the scope of substantive copyright protection, that law will be applied uniformly to foreign and domestic authors." *See Itar-Tass Russian Newspaper Agency v. Russian Kurier, Inc.*, 153 F.3d 82, 89 (2d Cir. 1998). Thus, Article 5(1) applies only to national *laws* granting *substantive* rights. A private settlement agreement is, by definition, not a law nor can it grant substantive statutory rights. As Congress made clear, only Congress can grant such rights. The Settlement does not provide or grant any substantive copyright rights; only Congress can do so. Instead, it is the settlement of copyright litigation between private parties that provides extensive remedies to members of the class.

Most importantly, the Settlement is careful to treat all covered books identically, without regard to whether their rightsholders are United States or foreign publishers or authors. Consistent with Section 104 of the Copyright Act, all copyright owners of books covered by the Settlement, regardless of their nationality, are entitled to exactly the same rights and receive exactly the same remedies under the Settlement.

2. Formalities and the Means of Redress

The Settlement and the class action opt-out procedure do not constitute a prohibited formality within the meaning of Article 5(2) of the Berne Convention.[17] The term "formality" is not defined in the Article. The World Intellectual Property Organization's guide to the Berne Convention provides some guidance. It states: "The word 'formality' must be understood in the sense of a condition which is necessary for the right to exist -- administrative obligations laid down by national law, which is not fulfilled, lead to loss of copyright." *See WIPO Guide to the Berne Convention for the Protection of Literary and Artistic Works* (Paris Act 1971) at 33 (1978). As the WIPO drafted the Berne Convention and is charged by the United Nations with administering it, its views should be entitled to great deference.

The process for reviewing and approving class action settlements under Rule 23 (b)(3) of the Federal Rules of Civil Procedure necessarily includes a mechanism to allow class members to opt out. That requirement of federal law is not a formality. It cannot lead to a loss of statutory copyright protection under the Copyright Act. Nor is the Settlement itself or the opt-out procedure of Rule 23 (b)(3) an administrative obligation laid down by national law that is necessary for the copyright right to exist. The Rule constitutes, instead, a *procedural* device.

The language of Article 5(2) makes clear that such procedural mechanisms – which are the means by which copyright infringement suits may be brought and resolved – fall exclusively within the province of national law and, as such, can be adopted by Congress and applied by federal courts consistently with the United States' obligations. *See* Article 5(2) (providing that "the extent of protection, as well as the *means of redress* afforded to the author to protect his rights, shall be governed *exclusively* by the laws of the country where protection is claimed [emphasis supplied]).

C. The Settlement is not "Legislating"

Some have expressed concern that the Settlement constitutes "judicial legislating" or that approval of this particular class action settlement is not otherwise properly within the province of the federal courts. That is not the case. Whatever one's view of judicial activism, a class action settlement negotiated by parties to a lawsuit and approved by a federal court only binds members of the class who have chosen not to opt out. It does not bind – or supplant the role of – Congress.

Moreover, settlements of litigation do not establish rights or obligations of general applicability, which is, by contrast, in the very nature of statutes. In this case, the Settlement only pertains to the rights of class members vis-à-vis Google, which is obtaining a non-exclusive license from rightsholders who have chosen to remain in the Settlement.

Furthermore, the class action device is widely used in all manner of state and federal common law and statutory claims. It has also been used in the context of federal copyright claims. Such actions have been settled, and those settlements are subject to the requirements of the Federal Rules and to judicial decisions interpreting such Rules. Congress has not chosen to exempt copyright infringement actions from the Federal Rules that authorize the use of the class action device in appropriate circumstances. Parties utilizing a congressionally created procedural device, and courts approving class action settlements, are, therefore, fully adhering to federal law.

Finally, nothing in the Settlement prevents Congress from legislating in the future as broadly as it wishes. Congress could, for example, pass orphan works legislation, to enable users to make use of a work after a due diligence effort to identify and locate the copyright owner.

End Notes

[1] The Guild had its beginnings as the Authors League of America, which was founded in 1912 by a group of book authors (including Theodore Roosevelt, who served as the League's founding vice president), short story writers, freelance journalists and a smattering of dramatists. In the 1920s, the Authors League broke into two groups: the Authors Guild and the Dramatists Guild of America.

[2] Pearl S. Buck (1938) (who served as Authors Guild president), William Faulkner (1949), John Steinbeck (1962), and Isaac Bashevis Singer (1978). One Guild member, Elie Wiesel (1986), has won the Nobel Peace Prize.

[3] Here's the math: we expect the settlement to make at least 10 million out-of-print books available, which, at an average of 300 pages per book, represents at least 3 billion pages of professionally written, professionally edited text. 20% of that is 600 million pages of text available at every desktop computer in the U.S. as a free preview. (For comparison, Encyclopedia Britannica is about 44,000 pages in print form; Wikipedia's featured articles total about 5,000 pages. All English Wikipedia articles, including stubs, total perhaps 3 million pages.)

[4] The impediments to making systematic use of the digital rights in out-of-print books are many, but they fall into two broad categories:

A. Rights clearance issues. Here, there are three different obstacles. First, it may be unclear who controls the rights because no one knows whether rights to a particular out-of-print book have reverted to the author or not (most standard trade book contracts ask an author to demand a reversion of rights before the contract formally terminates and the rights revert to the author). Second, it may be unclear who controls the digital rights for a particular out-of-print book, since many (but not all) older contracts make no mention of digital or analogous rights. Third, it may be difficult to find the author or publisher who controls the rights, particularly for older works. This is the so-called orphan works problem. (More on this later.)

B. Digitization and presentation issues. Here, the obstacles are capital and technological sophistication.

The rights clearance issues are largely addressed in Attachment A to the settlement, which deals with author-publisher issues and was the result of laborious negotiations between author and publisher representatives, and through the operation of class-action law.

The digitization and presentation issues have been handled with money and clever technology. Estimates of Google' s costs in scanning, digitizing and building the technical infrastructure to support the display of millions of out-of-print books run to about a billion dollars. (We have no special knowledge of Google' s costs.) Part of the challenge for Google was to find an efficient way to scan library books without damaging them by flattening them onto the scanner' s surface. It solved this through a patented technology that corrects for the distortions caused by the scanning of a book page that curves away from the scanner' s surface towards the book' s binding.

[5] This is because, for books published between 1923 and 1963, authors had to file renewal registrations to prevent their works from falling into the public domain. More importantly, the number of titles produced by the book publishing industry was far lower through most of the 20th Century than it was in its concluding decades. The median age of a copyright-protected book in a U.S. library is far younger than many commentators assume.

[6] Publishers are sought out too, of course. In some systems, such as in the U.K., different societies represent authors and publishers. In others, such as in Australia and Canada, the same society represents both authors and publishers.

[7] Capitalized terms used have the same meaning as in the Settlement Agreement.

[8] A " Copyright Interest" refers to a United States copyright interest to the extent implicated by a use covered by the Settlement Agreement, i.e., reproduction of a Book and the Display Uses and Non-Display Uses (described below). SA § 1.38. A " Book" refers to a written or printed, bound work that was published or distributed prior to January 6, 2009 and, if a " United States work" under 17 U.S.C. § 101, registered with the United States Copyright Office. The Settlement excludes from the definition of Book works that are Periodicals, unbound personal papers, and works in the public domain (including government works). SA § 1.16. An " Insert" refers to copyrighted textual (but not pictorial) content in a book whose rightsholder is different from the book' s rightsholder. SA § 1.72. *See* the sections of the SA cited in this footnote for more details concerning these definitions.

[9] Also, subject to agreement with the Registry, Google may, in the future, be authorized to exploit works in other revenue models, such as consumer subscriptions, print on demand, custom publishing, .pdf downloads, and summaries, abstracts and compilations. SA § 4.7.

[10] Exercise of the removal right does not vitiate Rightsholders' eligibility for the Cash Payment, described below.

[11] " Principal Work" refers to a book's principal written work (such as a novel, or a collection of short stories). SA §1.111. " Entire Inserts" are complete works, such as short stories, forewords and poems. " Partial Inserts" are all other content that meet the definition of an Insert, such as quotes from other works. SA §§ 1.50, 1.100.

[12] Based on discovery taken by the author plaintiffs, most form book publishing contracts in the late 1980s began to include express electronic rights grants to the publisher. Counsel for the authors and publishers thus agreed that authors would receive a greater revenue split for out of print Books published prior to 1987 than those published thereafter.

[13] The United States has entered into other international agreements that incorporate the United States' Berne Convention obligations by reference, such as the 1994 GATT TRIPS agreement. However, because these other agreements are derivative of the Berne obligations, we do not discuss them separately.

[14] 102 Stat. 2853-2861.

[15] Section 2 of the BCIA provides:

(1) The Convention for the Protection of Literary and Artistic Works, signed at Berne, Switzerland, on September 9, 1886, and all acts, protocols, and revisions thereto (hereafter in this Act referred to as the "Berne Convention") are not self-executing under the Constitution and laws of the United States.

(2) The obligations of the United States under the Berne Convention may be performed only pursuant to appropriate domestic law.

(3) The amendments made by this Act, together with the law as it exists on the date of the enactment of this Act, satisfy the obligations of the United States in adhering to the Berne Convention and no further rights or interests shall be recognized or created for that purpose.

Section 3(a) declares:

(a) Relationship with Domestic Law.--The provisions of the Berne Convention--

(1) shall be given effect under title 17, as amended by this Act, and any other relevant provision of Federal or State law, including the common law; and

(2) shall not be enforceable in any action brought pursuant to the provisions of the Berne Convention itself.

[16] Article 5(1) of the Berne Convention provides:

(1) Authors shall enjoy, in respect of works for which they are protected under this Convention, in countries of the Union other than the country of origin, the rights which their respective laws do now or may hereafter grant to their nationals, as well as the rights specially granted by this Convention.

[17] Article 5(1) of the Berne Convention provides:

The enjoyment and the exercise of these rights shall not be subject to any formality; such enjoyment and such exercise shall be independent of the existence of protection in the country of origin of the work. Consequently, apart from the provisions of this Convention, the extent of protection, as well as the means of redress afforded to the author to protect his rights, shall be governed exclusively by the laws of the country where protection is claimed.

In: Digital Books: Competition and Commerce
Editor: Oliver A. Hagen
ISBN: 978-1-61728-299-7

Chapter 4

STATEMENT OF MARYBETH PETERS, THE REGISTER OF COPYRIGHTS, BEFORE THE COMMITTEE ON THE JUDICIARY - HEARING ON COMPETITION AND COMMERCE IN DIGITAL BOOKS

Marybeth Peters

September 10, 2009

Hearing on Competition and Commerce in Digital Books: The Proposed Google Book Settlement

Chairman Conyers, Ranking Member Smith, and Members of the Subcommittee, I appreciate the opportunity to appear before you today to testify about the impact of the proposed Google Book Settlement Agreement on U.S. copyright law and policy as part of this hearing on Competition and Commerce in Digital Books.

SUMMARY

The Copyright Office has been following the Google Library Project since 2003 with great interest. We first learned about it when Google approached the Library of Congress, seeking to scan all of the Library's books. At that time, we advised the Library on the copyright issues relevant to mass scanning, and the Library offered Google the more limited ability to scan books that are in the public domain. An agreement did not come to fruition because Google could not accept the terms.

In 2005, we followed very closely the class action filed by The Authors Guild and its members and the infringement suit filed by book publishers shortly thereafter. The facts of the underlying lawsuits are simple. Google was reproducing millions of protected books in their entirety, without permission of the copyright owners, through systematic scanning operations set up with large research libraries. Once scanned, the books were indexed electronically, allowing end-users to search by title and other bibliographic information. Google returned hits to its customers that included the option of browsing "snippets" (e.g. several lines of the book), except for public domain books, which could be viewed and downloaded in their entirety. Google's search engine is free to users, but the company collects substantial revenue from the advertising that appears on web pages, including those pages on which images of, and information from, copyrighted books appear. The lawsuits raised complex and sometimes competing legal questions, including questions about intermediate copying, future markets, book digitization goals and fair use. Members of the legal community and the public debated the issues vigorously and anticipated what a Court decision on the merits might look like.

When the parties announced last fall that they had reached a settlement in what was becoming a long and protracted litigation, our initial reaction was that this was a positive development. But as we met with the parties, conversed with lawyers, scholars and other experts, and began to absorb the many terms and conditions of the settlement— a process that took several months due to the length and complexity of the documents— we grew increasingly concerned. We realized that the settlement was not really a settlement at all, in as much as settlements resolve acts that have happened in the past and were at issue in the underlying infringement suits. Instead, the so-called settlement would create mechanisms by which Google could continue to scan with impunity, well into the future, and to our great surprise, create yet additional commercial products without the prior consent of rights holders. For example, the settlement allows Google to reproduce, display and distribute the books of copyright owners without prior consent, provided Google and the plaintiffs deem the works to be "out-of-print" through a definition negotiated by them for purposes of the settlement documents. Although Google is a commercial entity, acting for a primary purpose of commercial gain, the settlement absolves Google of the need to search for the rights holders or obtain their prior consent and provides a complete release from liability. In contrast to the scanning and snippets originally at issue, none of these new acts could be reasonably alleged to be fair use.

In the view of the Copyright Office, the settlement proposed by the parties would encroach on responsibility for copyright policy that traditionally has been the domain of Congress. The settlement is not merely a compromise of existing claims, or an agreement to compensate past copying and snippet display. Rather, it could affect the exclusive rights of millions of copyright owners, in the United States and abroad, with respect to their abilities to control new products and new markets, for years and years to come. We are greatly concerned by the parties' end run around legislative process and prerogatives, and we submit that this Committee should be equally concerned.

As outlined below, the Copyright Office also believes that some of the settlement terms have merit and should be encouraged under separate circumstances. For example, the creation of a rights registry for book authors, publishers and potential licensees is a positive development that could offer the copyright community, the technology sector and the public a framework for licensing works in digital form and collecting micro- payments in an efficient and cost-effective manner. Likewise, the promise to offer millions of titles through libraries in formats accessible by persons who are blind and print disabled is not only responsible and laudable, but should be the baseline practice for those who venture into digital publishing. The ability of copyright owners and technology companies to share advertising revenue and other potential income streams is a worthy and symbiotic business goal that makes a lot of sense when the terms are mutually determined. And the increased abilities of libraries to offer on-line access to books and other copyrighted works is a development that is both necessary and possible in the digital age. However, none of these possibilities should require Google to have immediate, unfettered, and risk-free access to the copyrighted works of other people. They are not a reason to throw out fundamental copyright principles; they are a pretext to do so.

In the testimony below, we will address three specific points. First, we will explain why allowing Google to continue to scan millions of books into the future, on a rolling schedule with no deadline, is tantamount to creating a private compulsory license through the judiciary. This is not to say that a compulsory license or collective license for book digitization projects may or may not be an interesting idea. Rather, our point is that such decisions are the domain of Congress and must be weighed openly and deliberately, and with a clear sense of both the beneficiaries and the public objective.

Second, we will explain why certain provisions of the proposed settlement dramatically compromise the legal rights of authors, publishers and other persons who own out-of-print books. Under copyright law, out-of-print works enjoy the same legal protection as in-print works.[1] To allow a commercial entity to sell such works without consent is an end-run around copyright law as we know it. Moreover, the settlement would inappropriately interfere with the on-going efforts of Congress to enact orphan works legislation in a manner that takes into account the concerns of all stakeholders as well as the United States' international obligations.

Finally, we will explain that foreign rights holders and foreign governments have raised concerns about the potential impact of the proposed settlement on their exclusive rights and national, digitization projects. The settlement, in its present form, presents a possibility that the United States will be subjected to diplomatic stress.

Factual and Procedural Background

The proposed settlement, announced by the parties on October 28, 2008, would resolve claims that stem from Google's highly publicized Google Library Project. It is currently pending before the Honorable Denny Chin, United States District Court, Southern District of New York. Settlement Agreement, *The Authors Guild, Inc. et al. v. Google Inc.*, No. 06-CV-8 136 (S.D.N.Y. Oct. 28, 2008). The proceeding combines the unresolved claims of authors and book publishers as initially filed in two underlying actions: *The Authors Guild, Inc. v. Google Inc.,* No. 05 Civ. 8136 (S.D.N.Y Sep. 20, 2005) (a class action filed by representative authors and the Guild) and *The McGraw-Hill Companies, Inc. v. Google Inc.*, No. 05 Civ. 8881 (S.D.N.Y Oct. 19, 2005) (an action filed on behalf of five publishing companies).

By way of background, as of 2008 Google had digitized about 7 million books and other materials obtained through agreements with library collections at the University of Michigan, Stanford University, Oxford University, Harvard University and the New York Public Library, among others.[2] At a hearing convened by the European Commission in Brussels on September 7, 2009, Google announced that it has now scanned approximately 10 million books. Of these, Google estimates that about 1.5 million of these works are in the public domain. Many more may be works that are protected by copyright but have no identifiable or locatable copyright owner.[3]

1. Judicial Compulsory License

Class action lawsuits typically seek compensation for a class of similarly-situated persons who have suffered harm, or will suffer harm imminently, due to the defendant's past acts. The proposed settlement in fact resolves Google's past conduct by requiring Google to pay at least $60 for each book and $15 for each insert that was digitized prior to the opt-out deadline.[4] Proposed Settlement Agreement at 61, ¶ 5.1(a). But the class is overbroad and the settlement terms do not stop here.

Under the proposed settlement, the parties have crafted a class that is not anchored to past or imminent scanning, but instead turns on the much broader question of whether a work was *published* by January 5, 2009. As defined, the class would allow Google to continue to scan entire libraries, for commercial gain, into the indefinite future. The settlement would bind authors, publishers, their heirs and successors to these rules, even though Google has not yet scanned, and may never scan, their works.

We do not know the parties' reasons for defining the class according to whether a book was or was not published by January 5, 2009, but the result is to give Google control of a body of works that is many times larger than the 7 million works that were originally at issue. As defined, the class would bring into the settlement tent not only works that were published in the United States, and are therefore directly subject to U.S. law, but works published in most other countries in the world that have treaty relations with the United States.[5] While no one really knows how many works would be affected, Dan Clancy, the Engineering Director for the Google Book Search project, has been quoted as estimating that there are between 80 and 100 million books in the world.[6] As a practical matter, this means that the settlement would create for Google a private structure that is very similar to a compulsory license, allowing it to continuously scan copyrighted books and "inserts."[7]

Compulsory licenses in the context of copyright law have traditionally been the domain of Congress.[8] They are scrutinized very strictly because by their nature they impinge upon the exclusive rights of copyright holders. A compulsory license (also known as a "statutory license") is "a codified licensing scheme whereby copyright owners are required to license their works to a specified class of users at a government- fixed price and under government-set terms and conditions." Satellite Home Viewer Extension Act: Hearing Before the S. Committee on the Judiciary,108th Cong. (2004) (statement of David O. Carson, General Counsel, U.S. Copyright Office) (May 12, 2004). "[C]ompulsory licensing . . . break[s] from the traditional copyright regime of individual contracts enforced in individual lawsuits." *See Cablevision Sys. Dev. Co. v. Motion Picture Ass 'n of Am., Inc.,* 836 F.2d 599, 608 (D.C. Cir. 1988) (describing limited license for cable operators under 17 U.S.C. § 111). By its nature, a compulsory license "is a limited exception to the copyright holder's exclusive right . . . As such, it must be construed narrowly. . . ." *Fame Publishing Co. v. Alabama Custom Tape, Inc.,* 507 F.2d 667, 670 (5th Cir. 1975) (referring to compulsory licenses in the Copyright Act of 1909). Congress is the proper forum to legislate compulsory licenses when they are found necessary. *See* Melville B. Nimmer & David Nimmer, *Nimmer on Copyright* § 1.07 (2009) (Congress has authority to grant exclusivity and "may properly invoke . . . [n]onexclusivity under a compulsory license"); *cf. Cablevision* at 602 (citing *Teleprompter Corp. v. Columbia Broadcasting Sys.,* 415 U.S. 394, 414 (1974) (stating that it was Congress's role to address the issue of secondary transmissions if the Copyright Act of 1909 was inadequate). Compulsory licenses are generally adopted by Congress only reluctantly, in the face of a marketplace failure. For example, Congress adopted the Section 111 cable compulsory license "to address a market imperfection" due to "transaction costs accompanying the usual scheme of private negotiation. . . ." *Cablevision at 602.* "Congress' broad purpose was thus to approximate ideal market conditions more closely . . . the compulsory license would allow the retransmission of signals for which cable systems would not negotiate because of high transaction costs." *Id.* at 603.

As a matter of copyright policy, courts should be reluctant to create or endorse settlements that come so close to encroaching on the legislative function. Congress generally adopts compulsory licenses only reluctantly in the face of a failure of the marketplace, after open and public deliberations that involve all affected stakeholders, and after ensuring that they are appropriately tailored. Here, no factors have been demonstrated that would justify creating a system akin to a compulsory license for Google – and only Google – to digitize books for an indefinite period of time.

At very least, a compulsory license for the systematic scanning of books on a mass scale is an interesting proposition that might merit Congressional consideration. As stated above, various compulsory licenses have been carefully crafted over the years after extensive deliberation and consideration of the viewpoints of all affected stakeholders, though none apply to books or text. Among the issues Congress would want to consider are the pros and cons of allowing copyright users, rather than copyright owners, to initiate the digitization of copyrighted works; the rate of compensation that should be paid to copyright owners; and whether the same license terms should apply to mass digitization activities undertaken for the public interest by non profit organizations such as libraries, and for profit purposes by commercial actors. Congress also would want to consider whether all books merit the same attention, or whether differences can be drawn from the date of publication, the type of publication, or such facts as whether the rights holder is likely to be alive or deceased. Congress would need to consider the treaty obligations that may apply.

2. The Sale of Copyrighted Books without Consent of Rights Holders

The Copyright Office strongly objects to the treatment of out-of-print works under the proposed settlement. The question of whether a work is in-print (generally, in circulation commercially) or out-of-print (generally, no longer commercially available) is completely inconsequential as to whether the work is entitled to copyright protection under the law.

The Google Book Settlement gives Google *carte blanche* permission to use out of-print works by operation of the default rules. If a work is out-of-print, Google need not obtain permission before incorporating it into new "book store" products. These include on-line displays (up to 20% of a work), full-text purchases, and subscription products for institutional subscribers and library patrons. There are mechanisms by which the rights holder may stop Google after the fact and prospectively collect royalties that are predetermined by the Book Rights Registry ("BRR"). In summary, the out-ofprint default rules would allow Google to operate under reverse principles of copyright law, and enjoy immunity from lawsuits, statutory damages, and actual damages.

The activities that prompted the plaintiffs to file suit against Google – the wholesale scanning of books, electronic indexing and snippet display – are activities as to which reasonable minds might differ when considering whether such activities are acts of infringement or are, for example, fair use. However, the same cannot be said of the new uses that the settlement agreement permits Google to make of out-of-print works. We do not believe that even Google has asserted that, in the absence of this class action settlement, it would be fair use to undertake the new activities that Google would enjoy risk-free as a result of the settlement. In essence, the proposed settlement would give Google a license to infringe first and ask questions later, under the imprimatur of the court.

We are not experts on the proper scope of class action settlements, but we do wonder whether, as a constitutional matter, a class action settlement could decide issues that were not properly before the Court as part of the case and controversy presented during the litigation.[9] At the very least, within the context of copyright litigation, the class action mechanism has been used sparingly in recent years and has never resulted in the broad adoption of a settlement permitting extensive future uses of copyrighted products that were not the subject of the original infringement action.[10] A class action settlement that permits new activities for years to come, and removes the judicial remedies of millions of authors and publishers that are otherwise afforded by the Copyright Act, seems to us to be an excessive exercise of judicial power. The default rules for out-of-print books are not a small issue in the settlement because the *substantial majority* of books covered are out-of-print works—millions and millions of books. To be clear, the Office does not dispute the goal of creating new markets for out-of-print books – copyright duration has always been longer than the first print-run of a book and it has always been obvious that works will come in and out of favor, and in and out of print, during the term of protection. But copyright law has always left it to the copyright owner to determine whether and how an out-of-print work should be exploited.

Apart from its interest in ensuring the proper application of law and policy, Congress should be particularly concerned about the settlement since it would interfere with the longstanding efforts of Congress and many other parties to address the issue of orphan works. The broad scope of the out-of-print provisions and the large class of copyright owners they would affect will dramatically impinge on the exclusive rights of authors, publishers, their heirs and successors. Such alteration should be undertaken by Congress if it is undertaken at all. Indeed, this Committee has already invested significant time in evaluating the orphan works problem and weighing possible solutions. That process is not over. The Google Book Settlement would frustrate the Committee's efforts and make it exceedingly difficult for Congress to move forward. A much more productive path would be for Google to engage with this Committee and with other stakeholders to discuss whether and to what degree a diligent search for the rights holder should be a precondition of a user receiving the benefits of orphan works legislation, or whether a solution that is more like a compulsory license may make sense for those engaged in mass scanning. Whatever, the outcome, Congress is much better situated than the judiciary to consider such important and far-reaching changes to the copyright system.

As a side note, the Copyright Office would like to underscore for the Committee that out-of-print works and orphan works are not coextensive. Orphan works are works that are protected by copyright but for which a potential user cannot identify or locate the copyright owner for the purpose of securing permission. They do not include works that are in the public domain; works for which a copyright owner is findable but refuses permission; or works for which no permission is necessary, i.e. the use is within the parameters of an exception or limitation such as fair use. Many out-of-print works have rights holders who are both identifiable and locatable through a search. In fact, the U.S. works covered by the proposed settlement would all be searchable, at a minimum, through Copyright Office records because the settlement includes U.S. works only *if* they are registered. Proposed Settlement at 3, 9, ¶¶ 1.16, 1.72. Certainly, rights information may not be current and there may be disputes about rights between publishers and authors. However, these are the realities of the copyright system and the reason that Congress, the EU and other foreign governments have been working on a solution, with all of the deliberation and fine tuning that is appropriate. Until there is a legislative solution, it is our strong view that Google should conduct itself according to the same options available to other users of copyrighted works: secure permission; forego the use; use the work subject to risk of liability; or use the work in accordance with fair use or another limitation or exception.

The Office also notes that while the BRR might well provide a place for rights owners to come forward with contact information, it is also likely to have the unfortunate effect of creating a *false* database of orphan works, because in practice any work that is not claimed will be deemed an orphan. Many rights holders of out-of-print books may fail or refuse to register with the BRR for very good reasons, whether due to lack of notice, disagreement with the Registry's mission or operations, fear (*e.g.* privacy concerns) or confusion. The fact that the rights holder is missing from the BRR may also mean that he has no interest in licensing his work.

3. International Concerns

We are troubled by the fact that the proposed settlement implicates so many foreign works even when they have not taken steps to enter the United States market. While it would be appropriate to allow foreign nationals to participate voluntarily in licensing programs that may be developed by the BRR or other collectives, they should not be automatically included in the terms of the settlement. Moreover, we are aware that some foreign governments have noted the possible impact of the proposed settlement on the exclusive rights of their citizens. Indeed, many foreign works have been digitized by Google and swept into the settlement because one copy was in an academic research library in the United States. As a matter of policy, foreign rights holders should not be swept into a class action settlement unknowingly, and they should retain exclusive control of their U.S. markets.

The settlement imposes a requirement that all "U.S. works" be registered with the Copyright Office. U.S. works are, in relevant part, works that are first published on U.S. soil or published simultaneously in the United States and a treaty partner. *See* 17 U.S.C. § 101. That the parties would apply a registration requirement in this manner comes as no surprise in and of itself, especially since the issue is pending before the Supreme Court in another case. *See Muchnick v. Thomson* (*In re Literary Works in Elec. Databases Copyright Litig.*), 509 F. 3d 116, 122 (2nd Cir. 2007), *cert. granted sub nom. Reed Elsevier Inc. v. Muchnick*, 129 S.Ct. 1523 (2009). But in our view, this rule should be applied to all works in the class, i.e. to the extent foreign works are implicated at all, they should have been published in the United States and registered with the U.S. Copyright Office.[11]

For the past few months, we have closely followed views of the proposed settlement as expressed by foreign governments, foreign authors and foreign publishers. We have read numerous press accounts[12] and spoken with foreign experts. We know that some foreign governments have suggested that the settlement could implicate certain international obligations of the United States.[13] As the Committee is aware, the governments of Germany and France have filed objections with the Court. Memorandum of Law in Opposition to the

Settlement Proposal on Behalf of the Federal Republic of Germany, *The Authors Guild, Inc., et al. v. Google Inc.*, No. 05 Civ. 8136 (DC) (S.D.N.Y. Aug. 31, 2009); Memorandum of Law in Opposition to the Settlement Proposal on Behalf of the French Republic, *The Authors Guild, Inc., et al. v. Google Inc.*, 05 Civ. 8136 (DC) (S.D.N.Y. Sep. 8, 2009). Numerous foreign authors and publishers have raised concerns as well, including concerns about navigating the settlement from a distance. Indeed, the inherent difficulties of doing business internationally is one reason that typical collective management organizations work through counterparts in foreign countries, making it easier and more efficient for rights holders to protect their works on foreign soil, in foreign languages, under foreign laws, and using foreign currencies.

Some foreign governments have raised questions about the compatibility of the proposed settlement with Article 5 of the Berne convention, which requires that copyright be made available to foreign authors on a no less favorable basis than to domestic authors,[14] and that the "enjoyment and exercise of these rights shall not be subject to any formality."[15] For example, the Federal Republic of Germany has asserted that "[T]he proposed settlement is contrary to both the Berne Convention and WCT." Memorandum of Law in Opposition to the Settlement Proposal on Behalf of the Federal Republic of Germany at 4.

For purposes of this hearing, we are not suggesting that international obligations of the United States are at issue or necessarily would be compromised. However, it is a cause for concern when foreign governments and other foreign stakeholders make these types of assertions.

Conclusion

Mr. Chairman, thank you again for inviting me here today to present my observations and concerns. The Copyright Office welcomes any questions that the Committee has about the copyright implications of this unprecedented settlement agreement. To summarize, it is our view that the proposed settlement inappropriately creates something similar to a compulsory license for works, unfairly alters the property interests of millions of rights holders of out-of-print works without any Congressional oversight, and has the capacity to create diplomatic stress for the United States. As always, we stand ready to assist you as the Committee considers the issues that are the subject of this hearing.

End Notes

[1] Under certain narrow circumstances, libraries and archives may make use of works that are in their last 20 years of copyright protection, provided that the use is for purposes of preservation, scholarship, or research and that the library or archives has first determined, on the basis of a reasonable investigation that certain conditions apply. *See* 17 U.S.C. §108(h)(i).

[2] Google Books Settlement Agreement, http://books.google.com/googlebooks/agreement (last visited Sept. 4, 2009).

[3] Google Books Settlement, orphan works, and foreign works, http://blog.librarylaw.com/ librarylaw/ 2009/04/google-book-settlement-orphan-works-and-foreign-works.html (last visited Sept. 4, 2009).

[4] The settlement also addresses and resolves other issues such as the conduct of libraries, but the Office will not address those provisions for purposes of this preliminary assessment of issues with the Proposed Settlement Agreement.

[5] The United States enjoys international copyright relations with all but a small number of countries. *See* U.S. Copyright Office, Circular 38a: International Copyright Relations of the United States (rev. July 2009) (available at http://www.copyright.gov/circs/circ38a.pdf (last visited Aug. 25, 2009)).

[6] *See* "The Audacity of the Google Book Search Settlement," Pamela Samuelson, http://www. huffingtonpost. com/pamela-samuelson/the-audacity-of-the-googl_b_255490.html (last visited Aug. 12, 2009).

[7] The term "insert" is broad. It includes (i) text, such as forewords, afterwords, prologues, epilogues, essays, poems, quotations, letters, song lyrics, or excerpts from other Books, Periodicals or other works; (ii) children's Book illustrations; (iii) music notation (*i.e.*, notes on a staff or tablature); and (iv) tables, charts and graphs. Proposed Settlement Agreement at 9, ¶ 1.72.

[8] *See, e.g.*, 17 U.S.C §§ 111, 112, 114, 115, 118 and 119.

[9] As Judge Friendly stated in *National Super Spuds, Inc. v. New York Mercantile Exchange*, 660 F.2d 9, 18 (2d Cir. 1981), "If a judgment after trial cannot extinguish claims not asserted in the class action complaint, a judgment approving a settlement in such an action ordinarily should not be able to do so either." In *National Super Spuds*, a settlement purported to release the claims of class members who held both liquidated and unliquidated contracts when the original complaint only concerned persons who held liquidated contracts during a specific period of time. The Court held that the harm done by the unclear release of parties outweighed the benefits of settlement and reversed the settlement approval. *Id.*

[10] One of these class actions, *In re Literary Works in Electronic Databases Copyright Litigation*, MDL No. 1379 (S.D.N.Y.), is the remedies phase of an infringement suit brought by members of the National Writers Union, in which the writer-plaintiffs successfully challenged the sale of their newspaper and magazine articles in commercial databases. *See New York Times v. Tasini*, 533 U.S. 483 (2001). A settlement agreement has been proposed by the parties to the consolidated cases.[10] However, the proposed settlement, if finally adopted, would speak only to the activities originally at issue in the suit: the reproduction, display and distribution of copyrighted articles in electronic databases. Settlement Agreement, *In re Literary Works* (2005), ¶ 1(f). In contrast with the proposed settlement agreement, the *In Re Literary Works* settlement does not authorize the publisher and database defendants to further copy, package, and sell the copyrighted articles as part of new products such as subscriptions, books, or compilations, for example. Nor does it lock in licensing terms, including payment, for future kinds of activity.

[11] Article 5.1 of the Berne Convention provides for national treatment of authors by requiring that authors enjoy, in other Union countries, the rights provided to nationals of such Union countries. Berne Convention, 102 Stat. 2853 (1988). TRIPS also provides for national treatment in article 3.1; it requires Members to "accord to nationals of other Members treatment no less favorable than that it accords to its own nationals with regard to the protection of intellectual property." TRIPS Agreement, art. 3.1, 33 I.L.M. 81 (1994).

[12] *See, e.g.*, *Google Books Leaves Japan in Legal Limbo*, The Japan Times Online; *Germany Wants EU to Fight Google Books Project*, The Local, June 2, 2009 (quoting Foreign Minister Frank-Walter Steinmeiser); *Politicians Back Heidelberg Appeal: German Authors Outraged at Google Book Search*, Spiegel Online, Apr. 27, 2009 ("German politicians have voiced their support for an appeal by 1,300 German authors...known as the Heidelberg Appeal"—sent last week to German President Horst Kohler, Chancellor Angela Merkel and the heads of Germany's 16 federal states); Letter to the European Commission from the Federation of European Publishers and Presidents of National Publisher Associations, June 16, 2009 (available at http://www.danskeforlag.dk/download/pdf/323absb035.pdf(last visited Aug. 26, 2009)); Federal Ministry of Justice, *Zypries urges European action against Google Books, Press Release of the German Minister of Justice*; ("In Brussels today, Federal Minister Zypries stressed that...Brussels must take further steps that may be necessary to protect rights holders."); "*EU to study how Google Books impact authors*, Reuters, May 28, 2009 ("'The commission will carefully study the whole issue and, if need be, to take steps,'" Vladimir Tosovsky industry minister for the Czech EU presidency, told a news conference."); *Agreement concerning Google Book Search is a Trojan Horse*, Boersenverin des Deutschen Buchhandels, Nov. 11, 2008 ("[T]he American precedent model is out of the question for Europe...Germany and Europe have already implemented legal provisions and models which allow wide access to digital content while respecting the rules of copyright").

[13] By way of background, the United States is a party to important copyright treaties and bilateral agreements which impose minimum obligations for copyright protection and enforcement, on the one hand, and confine the scope of permissible exceptions and limitations on exclusive rights, on the other hand. These include the Berne Convention for the Protection of Literary and Artistic Works (Paris 1971), the World Trade Organization Agreement on the Trade-Related Aspects of Intellectual Property Rights ("TRIPS"), and the World Intellectual Property Organization Copyright Treaty, as well as many bilateral agreements that address copyright issues. *See, e.g.,* US-Peru Trade Promotion Agreement, Dec. 14, 2007, 121 Stat. 1454. Under Berne,[13] copyright protection is afforded to works published in any country that is party to one of the copyright treaties and agreements to which the United States is a party or by any national of that country.

[14] Berne Convention art. 5(1).

[15] Berne Convention art. 5(2).

In: Digital Books: Competition and Commerce
Editor: Oliver A. Hagen ISBN: 978-1-61728-299-7

Chapter 5

Competition that Works: Why the Google Books Project is Good for Consumers and Competitors, Testimony of David A. Balto, before the House Judiciary Committee, Senior Fellow, Center for American Progress Action Fund, Hearing on "Competition and Commerce in Digital Books"

David A. Balto

Mr. Chairman, Ranking Member Smith, and other members of the Committee, I am David Balto, a Senior Fellow at the Center for American Progress Action Fund. I appreciate the privilege of testifying before you today on Competition and Commerce in Digital Books which focuses largely on the Google Books project and the proposed settlement of litigation between Google and authors and publishers. As many of you know, I had a long career as a trial attorney in the Antitrust Division of the Justice Department and as the policy director of the Federal Trade Commission, and I frequently represent consumer and public interest groups in antitrust and intellectual property matters and testimony before Congress.[1] Based on my extensive review of the proposed settlement and the filings in the litigation, I strongly believe that the settlement in the Google Books project litigation does not pose significant competitive concerns and should be approved.

I have two simple points to my testimony. The Google Books project is a remarkable transforming achievement that we should all recognize has tremendous potential to democratize information and knowledge. I do not think anyone can dispute that. Second, the competitive concerns raised about specific narrow provisions of the settlement are not a basis to reject the settlement.

One of the greatest achievements in the last several years is the Google Books project, which scanned millions of books, many of which were available in only a handful of the most preeminent research libraries. This project led to class action litigation brought by publishers and authors charging a violation of copyright laws. To resolve the litigation, the protagonists entered into a settlement, which created a Book Rights Registry to make sure authors are appropriately compensated for their works. This settlement is currently under review by a federal district court.

THE TREMENDOUS BENEFITS FROM THE GOOGLE BOOKS PROJECT

The Internet is a great device for creating new markets, democratizing knowledge, and increasing competition. Google Books takes full advantage of this opportunity to expand the world's access to knowledge. Anyone can simply go on the web and, through Google Books, reach an almost endless array of information on nearly any topic. At the start of the 20th century, Andrew Carnegie spent an enormous sum to build the first truly public libraries in this country—before then, our libraries were for the most part only available to the educated and affluent. Google has taken on tremendous risk and expense to perform a comparable service, one that creates a virtual library of unprecedented proportions to millions of people, regardless of location, economic status, or resources. Thanks to the Google Books project, any individual anywhere in the United States will have access to an unprecedented corpus of information.

Under the settlement, Google Books will put out-of-print, expensive, or otherwise rare and hard-to-come-by texts online at reasonable prices. Libraries across the country will give people free access to Google Books: Google has pledged to give all public and university libraries across the country free, full-text, online viewing of millions of books at designated computers. Low-income, isolated, and underserved communities will have access to books that would have previously required enormous effort to track down. Lateef Mtima, a law professor at Howard University School of Law, has said that the settlement, by giving broad access to books and research materials to low-income communities, will "bridge the digital divide." Google's efforts will have the effect of expanding the body of knowledge readily available to the public, and make it even more accessible to people with disabilities and low-income or isolated populations.

It is not surprising that numerous public interest organizations and consumer rights advocacy groups have come out in favor of the settlement, including:

- United States Students Association
- American Association of People with Disabilities
- League of United Latin American Citizens
- Leadership Conference on Civil Rights
- National Federation of the Blind

Just to give one example of the remarkable consumer benefits, just consider the world of people with visual impairments—approximately 15-30 million in the United States alone. Google has committed to providing access to these users so that they "have a substantially similar user experience as users without print disabilities."[2] Google' s commitment to making all books available to people with visual impairments through screen enlargement, screen reader, and Braille display technologies will completely transform the educational experience for the blind and visually impaired in this country and, indeed, around the world. The National Federation of the Blind, the nation's leading advocate for access to information by the blind, has stated that the settlement will have "a profound and positive impact on the ability of blind people to access the printed word."[3]

The settlement will also provide researchers with the ability to analyze books and language in ways that were previously impossible. They will, for example, be able to search the entire digital library corpus to compare language and cultural development, and to track literary developments across countries. The potential to unlock knowledge is seemingly unlimited. Not surprisingly, universities around the country have overwhelmingly acknowledged these benefits. According to Michael Keller, Stanford University's librarian and publisher of the Stanford University Press, "[t]he settlement promises to change profoundly the level of access that may be afforded to the printed cultural record, so much of which is presently available to those who are able to visit one of the world's great libraries."[4]

As a public interest attorney, I represent many who often cannot afford to purchase books and who do not live near Harvard or Stanford or other research libraries. For all of these people, and millions more, the settlement will unleash greater access to a tremendous amount of information. The public benefit of the settlement is, to me, unmistakable.

Imagine if you are an energetic and inquisitive student at the Wayne State University, or at a community college in Alaska, and you are doing research on an obscure medieval text. Currently, you might have to travel to one of the great research libraries in Cambridge or Stanford to conduct research. Now, the corpus of many of the major research libraries will be only a click away on your computer.

The Google Books project is a tremendous achievement. The scope of that achievement and its potential to open access and information for millions of consumers should be considered as the committee and the courts consider the limited competitive issues raised by the settlement.

Competitive Concerns of the Settlement Are Unfounded

A tremendous amount of ink and paper has been spent raising the specter that the Google Books settlement will be harmful to consumers. In fact, some of the witnesses testifying today have claimed that Google will become a "cartel ringmaster" and coordinate a cartel of publishers and authors, a "monopolist" and charge excessive prices to consumers for digital books, and a "monopsonist" and decrease compensation to authors, thus resulting in the reduction of output of books. As I explain in a moment, none of these labels are supported by real world facts.

As I explain in my testimony it is easy, but misleading, to confuse popularity with market power (*i.e.*, the ability to harm consumers by raising price). Google may have popular products, but it has not harmed consumers. To paraphrase Senator Lloyd Bentsen in the Bentsen-Quayle vice-presidential debate in 1988: "I know monopolists, I have sued monopolists, and Google is no monopolist."[5]

My experience of over 15 years as an antitrust enforcer taught me to be cautious about substituting labels for real antitrust analysis. Antitrust labels may be attractive to the press or in a public debate, but they are not a substitute for analysis. As Justice Byron White cautioned in the *Broadcast Music* case thirty years ago, "easy labels do not always supply ready answers" to the question of whether conduct is illegal under the antitrust laws. Anyone looking at the unique issues in the settlement should be guided by that cautionary instruction. Rather than relying on easy labels, the critical question in analyzing whether conduct is anticompetitive is what is the *incentive and ability* for a firm to engage in the alleged anticompetitive conduct?

To answer that question we need look no further than Google's conduct in the numerous markets it currently participates in. Certainly, Google is large, innovative, and remarkably popular in search. But one should not confuse size and popularity with market power. Google does not charge consumers for the use of its very popular search product. That is because it lacks the incentive to charge for its use—its product is valuable to advertisers because of the numbers of users of the product. It also lacks the ability to charge because of the numerous free search alternatives also in the market. When it comes to allegations of anticompetitive conduct concerning Google search, the proof is in the pudding: Google lacks the incentive and ability to harm consumers.

So let's return to the claims of the critics of the settlement. Does Google, through the Book Rights Registry, have the incentive and ability to act as a "cartel ringmaster" directing thousands of authors or publishers in some sort of cartel waltz and facilitating a price increase? No. Over 95 percent of Google's revenue is through the sale of search advertising. It has no incentive to artificially increase book prices since that will undermine its advertising revenue. The amount of revenue Google would earn from trying to form or facilitate a publisher or author cartel is inconsequential compared to the advertising revenue they would lose. "Cartel ringmaster" is a job Google would never apply for.

The critics of the settlement create the specter of harm by conflating Google and the Book Rights Registry as if they were one entity. They are not. The board of the registry will be appointed by authors and publishers. The registry will be charged with getting the most money for authors, whereas Google will be interested in books being as inexpensive as possible, because it wants to increase usage of its site and search tools, so it can make money the way it usually makes money: advertising. Rather than be in a collusive relationship, Google and the registry are more likely to be in an adversarial relationship.

Would Google act as a "monopolist" and drive up the costs of digital publications offered through the registry? Again, acting in this fashion would be contrary to its economic incentives. Moreover, since there are many other sources for the most valuable and popular books, it seems unlikely this effort to charge supracompetitive prices would be anticompetitive.

Those who suggest that Google can act as a "monopolist" also misunderstand how books will be priced. Critics point to the fact that some books will be priced through an algorithm. The settlement provides only an offer to sell rightsholders' books at certain prices, however, and authors may reject this offer. This is no different than a situation in which a distributor offers multiple suppliers the chance to sell through the distributor at a given price, and sellers can reject or accept the offer. Further, I believe that 80 percent of digital books will initially be priced under $10.[6] Over time, the algorithmic pricing must be designed to mimic a perfectly competitive market.[7] Moreover, authors have the choice to opt-out of these "pricing bins" and determine their own price.[8] The result of this is that if Google sought to set monopoly prices, each author would have an incentive to undercut the market price in order to make additional sales.

And of course, all of these books will have to compete with: (1) public domain books which are free and downloadable from the site, and (2) the fact that every public and university library in the country will be able to have a free public access service terminal.[9] So any monopoly rents that Google might try to secure will be undercut by the fact that you can get the same thing for free at your local public library.

For the same reason that magazines practically beg you to subscribe, offering a fraction of the newsstand price, Google is more likely to pursue a model of broad access rather than price gouging for its subscription product, because that increases the advertising base. And even if Google misunderstood this, it is nevertheless obligated under the settlement to price the subscription to ensure "the realization of broad access to the Books by the public, including institutions of higher education."[10]

Finally, there is the claim that Google will act as a "monopsonist." A monopsonist is a firm that has market power in purchasing and may use that power to drive down compensation for a service provider resulting in a reduction in output. I have testified before Congress on behalf of farmers, nurses, doctors, and other healthcare workers about monopsony concerns, and there are very serious monopsony concerns in their markets. Monopsony concerns can arise where a service or good provider has limited outlets to sell its service or good. The Book Rights Registry is just one of dozens of means for authors or publishers to reach the market. If the prices are set too low, authors and publishers have numerous alternatives. The claims that Google would have the incentive or ability to become a monopsonist are simply fanciful.

THE SETTLEMENT ENHANCES THE POTENTIAL FOR ENTRY[11]

The remaining competitive concerns are actually very narrow. There are two questions. First, does Google' s access to "orphan works" limit the ability of alternative digital libraries to arise and compete against Google? Second, does a most favored nation provision restrict entry?

Both of these questions focus on entry barriers, and it is essential to get the entry barrier question right. The question is not whether it is hard to enter the digital books arena—it obviously is not easy. The operative question is whether the settlement *increases* entry barriers for the market. And to that, the answer is unequivocally no. As a group of 30 of the most preeminent antitrust law professors has observed, the "settlement overcomes barriers to entry for Google, without raising them for any rival because every right the settlement gives to Google to digitize, display, or sell books is expressly nonexclusive."[12] Thus, any right secured by Google can be shared with any existing or potential rival of the Google Books project.

Actually, the settlement makes it easier for others to follow in Google's footsteps in trying to enter the digital library arena. The settlement offers the potential to increase output and choice by expanding reducing legal and logistical barriers to similar digital books projects.

First, by expanding the public domain and resolving the uncertainty of rights of millions of books, the settlement will decrease entry barriers. Many books that have already been scanned by Google actually belong in the public domain; however, their status is currently unclear. In order to determine the copyright status of books under the settlement, Google is using records that have been scanned, compiled, corrected, and disseminated by Carnegie Mellon, Project Gutenberg, the Distributed Proofreaders, and Google itself. In addition to expanding the public domain in this way, the settlement also creates a procedure for publishers and authors to determine who should own digital rights, and it is expected that this will result in the clarification of these rights for thousands of publications. All of these efforts significantly expand access and facilitate entry. One study projects that the settlement should identify and resolve rights issues for at least 80 percent of rightsholders.[13]

Second, the settlement will also facilitate entry in another way: through its creation of the Book Rights Registry, an independent, nonprofit organization. The registry will significantly enhance the ability of subsequent entities to pursue book digitization initiatives.

Third, the registry will represent the interests of authors and publishers and locate rightsholders who have been separated from their works. With rightsholders' permission, it will be able to license millions of books, among them the most commercially valuable works covered by the settlement, to third parties, including Google competitors, on terms that might disfavor or disadvantage Google. Licenses will be provided on a nonexclusive basis, which means that authors and publishers will have the ability to negotiate with the registry or separately with other digital book providers. The ability to negotiate with other entities makes it unlikely that the registry could impose unreasonable license fees or otherwise restrict the availability of electronic books.

Finally, any institution with a license will be able to share the works in the project with other entities, including potential entrants into the digital books arena.

Access to Orphan Works

Much of the remaining competitive concerns surround Google's supposed exclusive access to orphan works. Orphan works are books that "retain their copyright but for which the rightsholders are unknown or cannot be found."[14] The concern is that Google's so-called control over these works will limit the ability of other other book scanning projects to be successful. These claims are overstated.

This Provision Affects a Small Number of Books

First, there are a relatively limited number of orphan works. According to a careful economic study, orphan works probably make up less than 9 percent of the works at issue. Other estimates suggest there may be as few as 1.4 million works, of which Google may have scanned around 580,000.[15] Further, this number should decrease over time as the efforts of the Book Rights Registry to reach out to authors and publishers allow them to clarify the copyright status of a number of the books in questions.

It is hard to see how this small set of orphan works would be some kind of bottleneck to limit market entry. Moreover, these works may be "orphaned" for a reason—they may have gone out of print because other works were superior.

As soon as an author or publisher clarifies their rights to an orphan book, that book ceases to be an orphan. The rights holder can then contract directly with potential users, such as libraries, and Google will not enjoy any special advantages at this stage, either.

The Settlement Does Not Grant Google Exclusive Access to Orphan Books

Second, the settlement only provides Google with nonexclusive access to orphan books. The settlement does not give Google any advantages over second entrants for books which still have orphan status, and Google's current efforts in fact lower the barriers for second entrants. Any company that chooses to begin scanning orphan books will face fewer obstacles than Google has confronted. Critics claim that other firms would be unlikely to attempt a similar class action suit to obtain their own default licensing rights to orphan books. This is not necessarily the case. By increasing the size of the public domain, clarifying uncertain rights, and addressing innumerable complex legal issues, Google Books simultaneously provides a map and blazes trails through previously unchartered territories that others will be able to use for their own explorations.

The Settlement Creates Value for Orphan Works

Finally, it is important to recognize that the value of orphan works is currently zero—there is no demand for them.[16] If they already had value, they would be in print or otherwise available to consumers. The Google Books Settlement creates economic value for these books by creating a means for consumers to access them and exercise demand for them.

The combined effect of the settlement's rights clarification efforts and financial incentives will be to clarify the copyright status of many works. As a result, the number of orphan works will be substantially reduced. The settlement enhances the ability of companies other than Google to provide digital access to books including orphan works.

The So-called Most Favored Nation Clause

A careful analysis of the most favored nation provision also shows that it is unlikely to increase entry barriers or harm competition in any other fashion. The "most favored nation," or MFN, provision of the settlement is extremely limited. It prevents the registry from offering a more favorable deal to other entities and the provision lasts for only ten years. I have written extensively and testified about the competitive implications of MFNs. These clauses can promote consumer welfare by permitting first movers to recoup their investments in innovation. Conversely, MFNs can impede entry and adversely impact competition. The MFN clause in the settlement falls into the former category.

The So-called "MFN" Clause is Narrow

The MFN provision is probably the most misunderstood provision of the agreement. This provision will have a limited effect on the marketplace. Nothing prevents the registry from striking better deals with Google competitors.[17] Specifically, the registry can pursue better deals with Google' s competitors for all in-print works, which represent the vast majority of today's book sales. Google competitors could also wait to see what the top- selling books are and strike deals only with respect to those books. Competitors could surely function in this way, like radio stations that seem to succeed despite playing only the top 30 songs over and over again. Section 3.8(a) pertains only to unclaimed works, that is, truly orphaned works, which no one comes forward to claim.[18] It also only operates *if* the registry is later authorized to license unclaimed works that fall within the scope of the settlement. The Book Rights Registry does not have this authority, and it may never receive this authority within the ten years that Section 3.8(a) operates. Thus, even Professor Picker's paper appears to concede that when the settlement first goes into effect, this provision will have no effect whatsoever.[19]

Clearly, the parties recognized that they were negotiating in an environment where Congress was very likely to legislate—orphan works. In fact, Google continues to advocate for orphan works legislation. This provision ensures that if legislation, or perhaps some future class action settlement, provides the registry authority that it does not have now that Google will receive equal treatment (not preferred treatment) under the new regime.[20] If that never happens, then this provision will have little relevance. This point has been lost on many of the provision's critics.

This is therefore a provision that will have no impact when the settlement goes into effect, and may never have any impact, and even if it does have an impact, it will only affect books which no one has claimed—books that have the *least* expected economic value.

The So-called MFN Clause Has Efficiency Justifications

Setting aside how narrow this provision is, it is worth noting that non-discrimination provisions are not uncommon. They are often used to protect the investment of first- movers who anticipate that their efforts will reduce the costs for second-movers, who can then strike more advantageous deals by free-riding off the investment of the first-mover. In fact, copyright law itself serves the same purpose: it protects the economic investment of the first mover—the author—who might not write if everyone could then free-ride off his investment. That is an undesirable outcome for all involved, including consumers. Google has invested $34.5 million in helping to create the registry, and millions more in legal fees, and thus, has a legitimate economic interest in protecting the value of that investment from free-riding.

Conclusion

As Judge Learned Hand instructed over half a century ago, the antitrust laws are not intended to punish "superior skill, insight, and industry." When Google announced this project five years ago, book scanning technology was in its relative infancy and cost-prohibitive. At its own risk, Google developed its own scanning technology, negotiated numerous agreements with libraries, and navigated the uncertainty surrounding complex copyright issues. Its ability to do all of these things led to a virtual library that offers an unprecedented level of access to millions of consumers. The purpose of the antitrust laws is to open access and opportunities and that is precisely what the Book Rights Registry and the settlement does. Innovation should not be confused with monopoly power. The Google Books settlement is in the public interest and should be approved.

APPENDIX A. PUBLIC INTEREST ADVOCACY BY DAVID BALTO

Testimony on Behalf of a Number of the Leading Consumer Groups in the United States

- On behalf of the **Consumer Federation of America, Consumers Union,** and **U.S. Public Interest Research Group** before the House Small Business Committee regarding health insurer consolidation (October 25, 2007).
- On behalf of the **Consumer Federation of America** before the Senate Judiciary Committee, Subcommittee on Antitrust, Competition Policy, and Consumer Rights regarding the concentration in agriculture and an examination of the JBS/Swift acquisitions (May 7, 2008).
- On behalf of the **Consumer Federation of America** before the House Small Business Committee regarding small business competition policy (September 25, 2008).
- On behalf of the **Consumer Federation of America** before the House Judiciary Committee regarding competition in the package delivery industry (September 9, 2008).
- On behalf of the **American Antitrust Institute, the Consumer Federation of America, the National Association for the Self-Employed,** and the **U.S. Public Interest Research Group** before the Senate Judiciary Committee, Subcommittee on Antitrust, Competition Policy, and Consumer Rights regarding consolidation in the Pennsylvania health insurance industry (July 31, 2008).
- On behalf of the **Consumer Federation of America** before the Senate Judiciary Committee, Subcommittee on Antitrust, Competition Policy, and Consumer Rights regarding the competitive impact of the Ticketmaster/Live Nation merger (February 24, 2009).
- On behalf of the **Consumer Federation of America** before the Senate Judiciary Committee regarding the nomination of Christine Anne Varney as Assistant Attorney General for the Antitrust Division of the Department of Justice (March 10, 2009).

Other Congressional Testimony

- Before the Senate Judiciary Committee, Subcommittee on Antitrust, Competition Policy, and Consumer Rights regarding GPO contract negations for medical supplies and devices (Sept. 14, 2004).
- Before the Senate Judiciary Committee, Subcommittee on Antitrust, Competition Policy, and Consumer Rights regarding the competitive impact of the XM/Sirius merger (March 20, 2007).
- Before the Antitrust Taskforce of the House Judiciary Committee regarding the impact of antitrust laws and their impact on community pharmacies and their patients (October 18, 2007).
- Before the Consumer Protection, Product Safety, and Insurance Subcommittee of the Senate Committee on Commerce, Science, and Transportation regarding competition in the health care marketplace (July 16, 2009).

Amici Curiae Briefs

- On behalf of **AARP, the Consumer Federation of America, Consumers Union,** and **Families USA.** *J.B.D.L. Corp. v. Wyeth-Ayerst Lab., Inc.*, Nos. 05-3860/3988 (6th Cir., Nov. 2005).
- On behalf of **the American Antitrust Institute** and **the Consumer Federation of America**. *McKinsey v. Peace Health*, No. 05-35627 (9th Cir., Mar. 2007).
- On behalf of the **American Antitrust Institute, AARP,** the **Consumer Federation of America, Consumers Union**, and **Families USA**. *In re DDAVP DIRECT PURCHASER ANTITRUST LITIGATION*, No. 06-5525 (2nd Cir., May 25, 2007).
- On behalf of **the American Antitrust Institute,** the **Consumer Federation of America**, and the **Organization for Competitive Markets**. *Federal Trade Commission v. Whole Foods Market*, No. 07-5276 (D.C. Cir., Aug. 23, 2007).
- On behalf of the **Consumer Federation of America** and the **Medical Device Manufacturers Association**. *Masimo Corp. v. Tyco Healthcare Group, LP*, Nos. 07-55960/56017 (9th Cir., Dec. 27, 2007).
- On behalf of the **American Antitrust Institute**, the **Consumer Federation of America**, and the **Public Patent Foundation**. *Rambus, Inc. v. Federal Trade Commission*, Nos. 07-1086/1124 (D.C. Cir., June 11, 2008).
- On behalf of several consumer groups. *Tafas and Smithkline Beecham Corp. v. Dudas and U.S Patent and Trademark Office*, No. 08-1352 (Fed. Cir., July 29, 2008).
- On behalf of **AARP, Patients not Patents**, and **the Public Patent Foundation**. *Apotex Inc. v. Sanofi-Synthelabo Inc.*, No. 09-117 (U.S. Supreme Court, Aug. 27, 2009).

Other Public Interest Advocacy

- On behalf of **the American Antitrust Institute** in opposition to the ExpressScripts/Caremark merger (white paper to FTC, Feb. 2007).
- On behalf of the **Consumer Federation of America, Consumers Union**, and **U.S. PIRG** in opposition to the FTC's proposed Consent Order against Kmart for deceptive marketing of gift cards (March 2007).
- On behalf of the **Organization for Competitive Markets**, and other groups in opposition to the Premium Standard/Smithfield merger (May 2007).
- On behalf of the **American Antitrust Institute, the Consumer Federation of America**, and **the Public Patent Foundation** on support for the FTC's proposed Consent Order in the Negotiated Data Solutions matter (white paper to FTC, April 24, 2008).
- On behalf of the **American Antitrust Institute** on Express Scripts' proposed acquisition of Wellpoint's PBM business (May 11, 2009).
- On behalf of the **Consumer Federation of America, U.S. PIRG**, and **Consumers Union** on pharmaceutical patent settlements (June 6, 2009).
- On behalf of the **Consumer Federation of America, U.S. PIRG**, and the **National Legislative Association on Prescription Drug Prices** on transparency by pharmacy benefit managers (letter to Speaker Pelosi, August 20, 2009).

APPENDIX B. THE PUBLIC INTEREST AND THE GOOGLE BOOK SEARCH SETTLEMENT

By David Balto, Senior Fellow, Center for American Progress Action Fund (September 9, 2009 version)

In 2004, Google began working with large research libraries to digitize their book collections and to make the content searchable online.Not long after the project was announced, a collection of authors and publishers sued Google for copyright infringement.After almost three years of negotiations, Google and the plaintiffs announced in October 2008 that they had agreed to a proposed settlement. While some commentators have lauded the settlement, others have vociferously claimed that it poses competitive concerns and does not promote the public interest.

As an advocate for consumer interests and a former antitrust enforcer, I took great interest in this debate early on and started to study the settlement. Over the last few months, I have learned much about Google Book Search, the ensuing litigation, the settlement, and the settlement's competitive implications.In doing so, I have come to the firm conclusion that the competition criticisms of the settlement are unfounded.I believe the settlement is good for consumers and should be approved.

At the outset, it is important to recognize what Google and the plaintiffs have accomplished. When Google started the project, book scanning technology was in its relative infancy and cost-prohibitive for operations at scale; the company thus had to develop its own scanning technology to move forward with the project.[21] At the same time, Google had to negotiate numerous agreements with libraries to gain access to their books and had to secure other rights directly from publishers and authors.[22]An additional deterrent was the great uncertainty surrounding the ownership of digital book rights. Indeed, it became abundantly clear that Google had undertaken considerable economic risk when its Book Search program became the subject of a class action lawsuit.

In settling the litigation, the publishers, authors, and Google have pursued a sound and necessary approach to resolving a number of rights-sharing problems that, until now, have posed seemingly insurmountable hurdles to making books digitally available.For example, by creating a nonprofit organization, the Book Rights Registry ("BRR" or "Registry"), to represent the interests of authors and publishers and to locate rightsholders who have been separated from their works, the settlement will significantly enhance the ability of subsequent entities to commence book scanning initiatives. As such, the settlement should be viewed in light of what it provides for the general public— increased access to the world's written cultural heritage, particularly books that have long been out of print. The settlement is, in other words, output-enhancing and procompetitive.

In this article, I first describe the settlement's consumer benefits, and then examine the impact of the settlement on entry barriers. I explain how, rather than increasing entry barriers, the settlement significantly decreases such barriers for other entities. I then focus on the criticisms posed by Professor James Grimmelmann,[23] and explain how the settlement is ultimately procompetitive and beneficial to the market.

THE SETTLEMENT WILL CREATE SIGNIFICANT CONSUMER BENEFITS

The Library of Alexandria, which Grimmelmann mentions in his articles, was one of the largest libraries in the ancient world. Although it remains unclear to this day how the library was destroyed, the leading theory is that Julius Caesar set a fire that unintentionally burned the library down in 48 B.C. during the Alexandrian War. The ancient texts contained within the library were destroyed, and the knowledge contained within the books was lost forever.

This historical incident is noteworthy for two reasons: first, on a practical level, the creation of digital copies of the world's books ensures that this disaster of the ancient world will not be repeated in modern times. Indeed, it would be borderline negligent *not* to use the electronic advancements of our generation to preserve the history and knowledge contained in books that have typically existed only on flimsy decaying paper. Second, despite the fact that there has been no fire or other calamity, a substantial percentage of the books in the United States (which Grimmelmann estimates is likely to be more than half of *all* books)[24] have been largely unavailable to the public, which amounts to a tragedy no less significant than the destruction of the Library of Alexandria. These are out-of-print books, which can be found in some public libraries and occasionally purchased second-hand, but which are otherwise inaccessible. There are a number of complications surrounding these books that have prevented them from being made more accessible. For many works, it is unclear whether they have fallen into the public domain because the rightsholder did not renew their copyright, as required by prior iterations of the Copyright Act. For some works, there is uncertainty as to who presently owns the rights to a book because the rights somehow got lost or otherwise disappeared over time. There is also ambiguity as to whether the author or the publisher owns the digital rights to hundreds of thousands of books. The combined impact of the varying shades of rights uncertainty has been to create gridlock,[25] which has prevented people from effectively accessing, and benefiting from, the knowledge encapsulated within an enormous number of books.

By settling their litigation, Google and the plaintiffs have ended the impasse. The settlement unleashes greater access to books, particularly to out-of-print books, which increases output in the marketplace of ideas.[26] Moreover, the settlement will serve as an equalizing force across socioeconomic, geographic, and linguistic barriers. Scholars and historians at the smallest schools in remote corners of this country will obtain the same access to knowledge as those at large, well-funded universities in our biggest cities. Citizens in poor communities will likewise have similar access to knowledge as those in affluent communities. And language barriers will be diminished under the settlement as Google's translation technology enables digital works in one language to be instantly translated into others.[27]

In addition to tearing down socio-economic and linguistic access barriers, the settlement will provide considerable benefits to those who are blind or otherwise print-disabled. Under the terms of the settlement, Google can provide books to "users with print disabilities so that such users have a substantially similar user experience as users without print disabilities."[28] The National Federation of the Blind, the nation's leading advocate for access to information by the blind, has stated that the settlement will have "a profound and positive impact on the ability of blind people to access the printed word."[29] The settlement will also provide researchers with the ability to analyze books and language in ways that were previously impossible. They will, for example, be able to search the entire digital library corpus to compare language and cultural development, and to track literary developments across countries. The potential to unlock knowledge is seemingly unlimited.

Universities around the country have overwhelmingly acknowledged these benefits. According to Michael Keller, Stanford's university librarian and publisher of the Stanford University Press, "[t]he settlement promises to change profoundly the level of access that may be afforded to the printed cultural record, so much of which is presently available to those who are able to visit one of the world's great libraries."[30] Paul Courant, the Dean of Libraries at the University of Michigan, has likewise observed that Google Book Search will provide:

> Ubiquitous online access to a collection unparalleled in size and scope, preservation of the scholarly and cultural record embodied in the collections of great research libraries, new lines of research, and greatly expanded access to the world's printed work for persons with print disabilities.[31]

For all these reasons, it is difficult to exaggerate the benefits that consumers will gain from the settlement—and important not to overlook the fact that these vast benefits will disappear if the settlement's detractors succeed in derailing its approval. Even some of the settlement's most vocal detractors, including Grimmelmann, acknowledge these fundamental facts.[32]

The Settlement Substantially Decreases Entry Barriers

Grimmelmann argues that the settlement will create various significant, if not insurmountable, barriers to entry that will prevent other potential competitors from competing in book scanning and online book sales. There is no doubt that book scanning is a difficult space to enter. Companies such as Microsoft and Yahoo have entered, seemingly determined that it would not be profitable, and exited.[33] Despite the costs and challenges, however, there are a number of entities that have entered book scanning and sustained their efforts.[34] As Grimmelmann acknowledges, the non-profit Open Content Alliance currently scans public domain works, and Amazon.com has the institutional capacity to make books available digitally on a "huge scale."[35] Most importantly, any challenges that do exist for large-scale book scanning should not be conflated with the question of whether the settlement increases or decreases entry barriers. The following sections explain the ways in which the settlement decreases entry barriers for book scanning and online book sales.

The Settlement Expands the Public Domain

Google has scanned approximately seven million works,[36] one million of which come from the company's partner program[37] and another million of which were published prior to 1923 and therefore are clearly in the public domain.[38] The copyright status of segments of the remaining five million works are unclear. Many of the books will actually belong in the public domain because the rights owners did not renew their copyrights, as required under previous iterations of the Copyright Act. As Grimmelmann notes, more than 85% of works published between 1923 and 1978 were not renewed, as required by the Copyright Act, and therefore fell into the public domain.[39] The catch here, because the Copyright Office did not keep effective records during this period, is identifying which works were not renewed. Carnegie Mellon, Project Gutenberg, the Distributed Proofreaders, and Google have, however, combined to scan, compile, correct, and disseminate these records.[40] And Google is now using the records to determine the copyright status of books under the settlement.

Notably, the Copyright Office examined the feasibility of scanning these records and placing them online in 2006. They expressed reluctance to commit to such a project, however, because it would "involve a significant expenditure of resources" [41] as "preliminary figures estimated the costs to be about $35 million."[42] Fortunately, however, government inaction has not impeded progress because Google and others have filled the void – without the expenditure of public funds.

Each time that Google determines a book belongs in the public domain, consumers benefit by being able to download an entire PDF version of the text. Potential book scanning entrants benefit by having fewer books with uncertain rights — and can thereby avoid incurring expenses that Google had to incur to navigate these uncertain rights. By facilitating the clarification of the public domain status of potentially millions of works, the settlement thus significantly expands access and facilitates entry.

The Settlement Will Lead to the Resolution of Uncertain Digital Rights

Another significant source of uncertainty surrounding the status of many out-of-print works is the ownership of the digital rights associated with particular books. Up until the 1990s, publishing agreements typically did not allocate digital rights to books, and publishers and authors have disagreed about who should retain these rights by default. The resulting standoff has harmed authors and publishers alike by denying them the possibility of gaining revenue from the sale of digital works. Consumers are likewise harmed by the loss of an effective option for accessing these publications. The settlement resolves this problem by creating a procedure for publishers and authors to determine who should own these digital rights. In addition, while publishers or authors may not have had any incentive to assert their digital rights in the past because their books were out of print or otherwise no longer profitable, the settlement creates an incentive for authors or publishers to come forward to assert their rights. Specifically, individuals and institutions can purchase access to out-of-print works that are still under copyright and the revenues that will be derived from these sales, combined with various inclusion fees described in the settlement, create financial incentives for owners of out-of-print books to claim them.

To ensure that copyright holders worldwide are aware of these financial incentives, the settlement has established the most comprehensive class-action notification program *ever*. Specifically, Google has funded a large direct-mail effort, created a dedicated Web site about the settlement in 36 languages, and spent about $7 million on advertising in newspapers, magazines, and even poetry journals, with at least one ad in each country.[43]

By creating an entity to resolve disputed digital rights claims between authors and publishers, providing financial incentives for rightsholders to claim their works, and funding the world's largest class-action notification program, the settlement generates substantial pro-competitive benefits and facilitates entry.

The 'Orphanage' Post-settlement

Grimmelmann alleges that the settlement will provide Google with "exclusive control" or a "monopoly" over "orphan works," which are works for which the current rightsholder is unknown. These claims lack economic substance.

First, Google will affirmatively *not* obtain a monopoly over orphan works because the settlement does nothing to make entry more difficult for a second entrant. Indeed, any company that chooses to begin scanning orphan works will face fewer obstacles – due to the settlement – than Google has confronted. In particular, by increasing the size of the public domain, clarifying uncertain rights, and addressing innumerable complex legal issues, the settlement simultaneously provides a map and blazes trails through previously unchartered territories than others will be able to use for their own explorations.

Second, we should remember that these works were orphaned because the rightsholders failed to keep track of their interests in the books. Thus, while it is obviously exciting that Google's scanning efforts could result in the (re)discovery of works of genius that have sat anonymously in the stacks of our nation's research libraries, we should also not forget that the true remaining orphans probably had indifferent parents and are likely of little value. As Roy

Blount, President of the Authors Guild, noted, orphan works are books that "have been deemed unfit for continued commerce by traditional print publishers." [44]

Third, the number of true orphan works that will exist post-settlement warrants closer scrutiny. Contrary to the claims of Grimmelmann and others, the number of orphan works will be substantially reduced. Specifically, we know that:

- Over 85% of copyrighted works prior to 1964 were not renewed and belong in the public domain, which could amount to as many as 1.5 million works.[45]
- The settlement creates financial incentives for copyright owners to come forward and claim works, which will facilitate rights clarification,[46] and
- The Registry is obliged to locate authors under the settlement and, according to Roy Blount, studies on efforts to locate the rightsholder of out-of-print work have typically resulted in an 80-85% success rate. [47]

It is thus overwhelmingly apparent that the combined effect of these rights clarification efforts and financial incentives will be to clarify the copyright status of hundreds of thousands, if not millions of works, which will be *an enormous improvement* over the status quo. Studies on the number of true orphan works in the United State have, moreover, found few fewer true orphans than critics of the settlement allege. Peter Hirtle, for example, concluded that there is a total of 12 million out-of-print works in the United States and that approximately 1.4 million could be true orphans.[48] If Google has scanned about 5 million out-of-print but in-copyright works, this means its library might include about *580,000 orphan works*.

According to Brewster Kahle,[49] the head of the Open Content Alliance, it costs Google about $10 to scan a book and costs the Internet Archive about $30 to scan a book with "superior quality."[50] It thus would cost a rival to Google somewhere between $5.8 million and $17.4 million to scan the same body of orphan works—an expensive undertaking to be sure, but one that is certainly achievable. As such, it is entirely disingenuous to claim that the settlement increases entry barriers surrounding orphan works or that Google will obtain a monopoly over them.

The Class-action Nature of the Settlement

Grimmelmann contends that the settlement is a good deal only for Google because "[i]n a post-settlement world ... potential competitors [in book scanning] face one insurmountable hurdle: copyright law."[51] This is absurd because the copyright law is no different in the "post-settlement world" than in the "pre-settlement world." Indeed, when Google started scanning these books, it was sued for alleged copyright violations. And as the settlement shows, copyright law is a *surmountable* hurdle and indeed, the settlement itself is a valuable guide for others seeking to surmount it.

Nonetheless, Grimmelmann posits that any subsequent entrant would be "sued into oblivion by a mob of angry copyright owners."[52] Conversely, he expresses concern that the class action nature of the settlement will serve as a "remarkably effective barrier to entry" because a subsequent entrant would not be sued by a class action, similarly to Google.[53]

These concerns are unfounded once one understands the requirements for class certification, which are found under Rule 23 of the Federal Rules of Civil Procedure. Rule 23 states that a class should be certified if it comprises numerous plaintiffs with common questions of law or fact that share typical claims and defenses and have interests that can be fairly and adequately protected by class representatives. Defendants usually bitterly contest certification because it can mean the difference between exposure to individual damage claims that might amount to hundreds or thousands of dollars and the nationwide aggregation of these claims that might amount to millions or even billions of dollars in damages. In the book scanning context, however, a defendant will benefit from, and therefore *support*, class certification because it enables the defendant to simultaneously resolve all its copyright issues. Given the concerns of copyright owners, a supportive defendant, and the *Authors Guild v. Google, Inc.* class certification precedent, it seems highly likely any prospective class will be certified.

Innovation and Transparency

Grimmelmann also worries about Google becoming a "chokepoint" for book distribution, and about the possibility of Google censoring information it does not want distributed. The irony of this argument is overwhelming. The Internet has facilitated an unprecedented explosion in the ability of individuals to publish content – regardless of how bizarre, tasteless, or offensive it may be – to the world at large. And Google, as the world's most successful search engine, facilitates greater access to published content than *any other entity in history.* The suggestion that the company thus might find the need to censor the content of books obtained from its library partners while simultaneously providing search results to users on virtually any topic is seemingly fanciful. In addition, Grimmelmann's argument appears to be premised on a Brave New World, where all other forms of distribution and access to literature have apparently become extinct. Even if Google were to decide not to distribute certain works through its system, one must remember that this settlement is non-exclusive, thus allowing copyright holders to freely negotiate with any potential competitors, and that these "censored" works could still be published, sold, and distributed through all the means of publication that exist today.

Furthermore, the settlement provides a mechanism to address this specific concern—if Google wishes to remove a book from its search results (which Grimmelmann notes is the company's First Amendment right), then it must inform the BRR and provide the BRR with a digitized copy of that work. No such mechanism would exist if Google were to have successfully litigated its case against the authors and the publishers, so the settlement actually provides *greater* transparency than the 'but for' world.

For all of the above reasons, there simply is no merit to the contention that the settlement will raise entry barriers for other would-be providers of digital books. The settlement will result in an expansion of the public domain and in the resolution of disputes over digital rights that are currently impeding our ability to access to vast quantities of books. And while issues inherent in our current system of copyright law, including orphaned works, will continue to impede access, none of these problems is increased by the settlement. In the post-settlement world, it will be in no way harder and in many ways easier for companies other than Google to provide digital access to books.

REMAINING ANTITRUST CONCERNS ARE EQUALLY UNFOUNDED

Grimmelmann raises a number of specific antitrust concerns about the settlement, including that it facilitates price-fixing, confers monopoly power on the Registry, and contains an anticompetitive most-favored nation (MFN) provision.[54] Although these are legitimate points to examine, a careful examination of the settlement and the underlying economics shows they are unfounded.

The Settlement Does Not Facilitate Price-fixing

Grimmelmann and others contend that the settlement will foster price-fixing among publishers and authors by creating an e-book program for consumer sales that gives copyright holders the option of allowing Google to determine sales prices. He points to a settlement provision that sets out twelve possible bin prices ranging from $1.99 to $29.99 and excitedly observes that "[s]urely the sophisticated companies with expert lawyers who drafted this settlement agreement wouldn't use it to set up a system of naked price- fixing ... ."[55]

When these consumer sales provisions are examined, it becomes readily apparent that the settlement does not establish a cartel. These provisions actually provide two ways by which prices for consumer book sales can be determined: (1) the individual author or publisher can specify price; or (2) Google can select a price based on its own algorithm.[56] For algorithmic pricing, which is the focal point of Grimmelmann's concern, the settlement states that "Google may change the price of an individual Book over time" and the "distribution of Books ... among the Pricing Bins may change over time."[57] These provisions also prevent the Registry, authors, or publishers from interfering with Google's pricing freedom. The settlement does not, in other words, create "a system of fixed prices." Grimmelmann is plainly wrong.

Courts and antitrust authorities have recognized that in many environments collective price setting can be procompetitive when intellectual property is involved. The Supreme Court has noted that, especially when dealing with copyright issues, price-fixing is "not a question simply of determining whether two or more potential competitors have literally 'fixed' a 'price.'"[58] The Supreme Court ruled in *Broadcast Music Inc. v. CBS* ("BMI") that music copyright clearinghouses were not violating the antitrust laws *per se*, and in doing so, it expounded upon the purpose of copyright in general that aptly applies here:

> Although the copyright laws confer no rights on copyright owners to fix prices among themselves or otherwise to violate the antitrust laws, we would not expect that any market arrangements reasonably necessary to effectuate the rights that are granted would be deemed a *per se* violation of the Sherman Act. *Otherwise, the commerce anticipated by the Copyright Act and protected against restraint by the Sherman Act would not exist at all or would exist only as a pale reminder of what Congress envisioned.*[59]

As in *BMI*, the Google settlement will not facilitate a *per se* cartel violation, but instead will accomplish "the integration of sales, monitoring, and enforcement against unauthorized copyright use."[60] This integrated facilitation of copyright enforcement and market enhancement in this manner is, as Grimmelmann himself has so often noted, unabashedly procompetitive and beneficial.

The Registry is Not a Potential Monopolist

Grimmelmann contends that the "Registry is also a potentially dangerous monopoly" because it "will speak on behalf of an entire industry."[61] At the same time, he argues that the "best way to make the Registry work well" is to expand its scope and powers.[62] This is a perplexing contradiction. No economist or antitrust lawyer would recommend expanding the powers of an entity to eliminate the monopolization concerns it poses. If the Registry were a "dangerous monopoly," it would be appropriate to curtail its powers. Period.

But his competitive concerns are simply misplaced. The Registry is no different than other collective rights management organizations that have been acknowledged as procompetitive under the antitrust laws. For example, the Department of Justice ("DOJ") previously examined a licensing program proposed by the Copyright Clearance Center, Inc. ("CCC"), a non-profit organization created by authors, publishers, and users of copyrighted material to facilitate copyright licensing and clearance.[63] The DOJ concluded that CCC's licensing arrangement, which authorizes users to make unlimited copies of any work in CCC's library for an annual fee, did not raise competitive issues in part because the copyright holders that are CCC members may continue to negotiate separate licensing arrangements with users and competing firms.[64] As part of its analysis, the DOJ also noted the procompetitive benefits of the blanket licensing agreement, such as encouraging use of copyrighted works among a wider range of users.[65] Similar to the CCC, the Registry will offer blanket licenses to digital book users on a nonexclusive basis. This is a key provision that Grimmelmann notes, but seemingly ignores its full implications.[66] Non-exclusivity means that, as with the CCC, authors and publishers will have the ability to negotiate with the Registry or separately with digital book providers. And because authors and publishers will have the ability to negotiate with other entities, the DOJ's conclusion regarding this provision in the settlement should be the same as the DOJ's conclusion with respect to the CCC—that absent exclusivity, it is unlikely that the Registry could impose unreasonable license fees or otherwise restrict the availability of electronic books.

Just because the Registry is important does not mean it is a monopoly, but ultimately it will provide significant benefits to consumers. Under the Settlement, the Registry will own and maintain a rights information database for books and their authors and publishers, locate rights holders, distribute payments from Google to rights owners, and assist in the resolution of disputes between those claiming to hold digital rights.[67] The Registry thus has the potential to significantly increase the amount of information and the number of books available in digital form to consumers.

The Most-favored Nation Clause is Procompetitive

Some critics suggest that the settlement's so-called Most-Favored Nation clause ("MFN") raises concerns; Grimmelmann suggests that it is "[t]he most pressing problem" because it "explicitly guarantees Google a privileged position."[68] I have written extensively and testified about the competitive implications of MFNs.[69] These clauses can promote consumer welfare by permitting first movers to recoup their investments in innovation. Conversely, MFNs can impede entry and adversely impact competition. The MFN clause in the settlement falls into the former category.

The MFN only applies in limited circumstances. Specifically, provision 3.8(a) states that the Registry will not license to third parties on terms that "disfavor or disadvantage Google" when such authorizations include rights granted from a "significant portion" of unclaimed works.[70] In other words, the Registry can license all the claimed works and some of the unclaimed works to third parties on terms that disfavor or disadvantage Google. The remainder of the works can moreover be licensed on identical terms to Google. It seems unlikely the MFN will hinder competition.

Conclusion

The universally accessible, searchable, digital library that will be realized by the Google Book Search settlement will provide unprecedented benefits to consumers worldwide. The settlement is an efficient and socially beneficial solution to the significant rights uncertainty that currently surrounds many books. Many of the leading critics of the settlement, such as James Grimmelmann, have failed to appreciate these procompetitive benefits while also dramatically overstating the antitrust risks. From a consumer welfare perspective, the settlement should unquestionably be approved.

End Notes

[1] A listing of my advocacy for consumer groups such as the Consumer Federation of America, Consumers Union, Families USA, AARP, and US PIRG is attached. Many of the arguments discussed in my testimony are set forth in an earlier paper: David Balto, "The Earth Is Not Flat: the Public Interest and the Google Book Search Settlement: A Reply to Grimmelmann," July 22, 2009, available at http://www.acslaw.org/node/13812. A later version of the paper is attached.

[2] eSight Community News, "Google Settlement with Authors, Publishers," Press release, November 23, 2008, available at http://www.tabinc.org/blog/archives/2008/11/google_settleme.html.

[3] Ibid.

[4] University of Michigan News Service, "Major Universities See Promise in Google Book Search Settlement," Press release, October 28, 2008, available at http://www.ns.umich.edu/htdocs/releases/story.php?id=6807.

[5] Ed Black, president and CEO of the Computer and Communications Industry Association, made the identical observation in his June 10, 2009 article for *The Hill* entitled "Google venture; no violation of copyright or antitrust laws, saying I know what a high-tech monopoly looks like. This isn't one." See http://thehill.com/opinion/letters/7012-google-venture-no-violation-of-copyright-or-antitrust-laws

[6] See Settlement § 4.2(c)(ii).

[7] See Settlement § 4.2(b) ("The Pricing Algorithm shall base the Settlement Controlled Price of an *individual* Book upon aggregate data collected with respect to Books that are similar to such Book.") (emphasis supplied).

[8] See Settlement § 4.2(b)(i) ("A rightsholder may select one of two pricing options for Consumer Purchases").

[9] See Settlement § 4.8(a)(i)..

[10] See Settlement § 4.1(a)(i).

[11] In a very thoughtful amicus brief, the Computer and Communications Industry Association explains how the settlement will enhance market entry. They explain [b]y thus lowering risks and licensing costs, the settlement may encourage competitors to enter the market, or may encourage participants who exited the market to reenter. Brief of Amicus Curiae Computer and Communications Industry Association on Proposed Settlement at 11, *Authors Guild, Inc., et al. v. Google, Inc.*, No. 05-8136 (S.D.N.Y. September 8, 2009).

[12] Brief of Amicus Curiae *Antitrust and Economics Professors in Support of the Settlement* at 19, *Authors Guild, Inc., et al. v. Google, Inc.*, No. 05-8136 (S.D.N.Y. September 8, 2009).

[13] Ibid at 5

[14] Jerry A. Hausman and J. Gregory Sidak, Google and the Proper Antitrust Scrutiny of_Orphan Works, *Journal of Competition Law and Economics* 5 (3) (2009): 411, 419.

[15] Brief of Amicus Curiae Computer & Communications Industry Association on Proposed Settlement at 12, *Authors Guild, Inc., et al. v. Google, Inc.*, No. 05-8136 (S.D.N.Y. September 8, 2009); A. Mark Lemley, A New Balance Between IP and Antitrust, John M. Olin Program in Law and Economics. Stanford Law School. Working Paper No. 340. (April 2007)..

[16] Jerry A. Hausman and J. Gregory Sidak, "Google and the Proper Antitrust Scrutiny of_Orphan Works," *Journal of Competition Law and Economics* 5 (3) (2009): 411, 419.

[17] Settlement § 3.8(a).

[18] The provision is limited to rights granted from a significant portion of rightsholders other than registered rightsholders. Because registered rightsholders is defined by the settlement to mean any person who is a rightsholder and who has registered with the registry his, her or its copyright interest in a book or insert. See § 1.122, and a rightsholder is a member of the settlement class who does not opt out , see § 1.132, this section only governs works of settlement class members that do not register with the registry.

[19] Randall C. Picker, The Google Book Search Settlement: A New Orphan Works Monopoly? John M. Olin Law and Economics Working Paper No. 462, at 21.

[20] Antitrust scholars Areeda and Hovenkamp take the position that provisions requiring equal treatment are unlikely to cause antitrust problems. III.B Philip Areeda and Herbert Hovenkamp, Antitrust Law ¶ 768a6, at 159-160 (3rd edition 2008).

[21] Maureen Clements, The Secret of Google s Book Scanning Machine Revealed, National Public Radio, April 30, 2009, available at www.npr.org/blogs/library/2009/04/the_granting_of_patent_7508978.html.

[22] In December 2004, Google entered agreements with the libraries at Harvard University, the University of Michigan, Stanford University, Oxford University, and the New York Public Library to scan parts of their collections of books and make the contents searchable online. Google received access to the collections, while the libraries received an electronic copy of the books that they provided to Google. See Google Checks Out Library Books, Google Press Release, December, 14, 2004, available at www.google.com/press/pressrel/print_library.html.

[23] James Grimmelmann, The Google Book Search Settlement: Ends, Means, and the Future of Books (American Constitution Society, April 2009), available at http://www.acslaw.org/files/Grimmelmann%20Issue%20Brief.pdf; James Grimmelmann, How to Fix the Google Book Search Settlement, *Journal of Internet Law* 1 (April 2009), available at http://works.bepress.com/cgi/viewcontent.cgi?article=1022&context=james_grimmelmann.

[24] Grimmelmann, "The Google Book Search Settlement: Ends, Means, and the Future of Books,", supra note 3 at 8.

[25] See generally, Michael Heller, *The Gridlock Economy* (New York: 2008).

[26] James Gleick, "How to Publish Without Perishing, *The New York Times*, November 29, 2008, available at http://www.nytimes.com/2008/11/30/opinion/30gleick.html ("As a way through the impasse, the authors persuaded Google to do more than just scan the books for purposes of searching, but go further, by bringing them back to commercial life. Under the agreement these millions of out-of-print books return from limbo.... This means a new beginning a vast trove of books restored to the marketplace.").

[27] See for example CNET News, "Google's Digital-Book Future Hangs in the Balance," June 15, 2009, available at http://news.cnet.com/8301-1023_3-10262203-93.html (noting that Google increasingly sophisticated translation technology ... could bulldoze literary language barriers").

[28] eSight Community News, "Google Settlement with Authors, Publishers," Press release, November 23, 2008, available at www.tabinc.org/blog/archives/2008/11/google_settleme.html.

[29] Ibid.

[30] University of Michigan News Service, "Major Universities See Promise in Google Book Search Settlement," Press release, October 28, 2008, available at www.ns.umich.edu/htdocs/releases/story.php?id=6807.

[31] Paul Courant, "Other Voices: Google Agreement Will Extend U-M Libraries Accessibility," Mlive.com, June 24, 2009, available at www.mlive.com/opinion/ann-arbor/index.ssf/2009/06/other_voices_google _agreement.html.

[32] See Grimmelmann, "How to Fix the Google Book Search Settlement," ,supra note 3 at 12 ("Everyone is better off than they would be in a world without Google Book Search . . . . ").

[33] Grimmelmann, The Google Book Search Settlement: Ends, Means, and the Future of Books,", supra note 3 at 10.

[34] See, for example, Emory University, "Emory Partnership Breaks New Ground in Print-On-Demand Books Press release, June 6, 2007, available at www.emory.edu/news/Releases/KirtasPartnership1181162558.html (announcing partnership with Kirtas Technologies, Inc., a maker of digital scanning technology to apply automated scanning technology to thousands of rare, out-of-print books in its research collections)..

[35] Grimmelmann, "The Google Book Search Settlement: Ends, Means, and the Future of Books,", supra note 3 at 10..

[36] Juan Carlos Perez, "In Google Book Settlement Business Trumps Ideals," IDG News Service, October 30, 2008) available at www.pcworld.com/businesscenter/article/153085/in_google_book_settlement_business_trumps_ideals.html.

[37] Ibid.

[38] Ibid.

[39] Grimmelmann, "The Google Book Search Settlement: Ends, Means, and the Future of Books,", supra note 3 at 8.

[40] Google Book Search Blog, U.S. Copyright Renewal Records Available for Download, available at http://booksearch.blogspot.com/2008/06/us-copyright-renewal-records-available.html.

[41] United States Copyright Office, "Report on Orphan Works" (January, 2006), 29-30 n.46, available at www.copyright.gov/orphan/orphan-report.pdf.

[42] Ibid.

[43] Noam Cohen, "A Google Search of a Distinctly Retro Kind, *The New York Times*, ,March 3, 2009,, available at www.nytimes.com/2009/03/04/books/04google.html (noting that 200 advertisements have run in more than 70 languages: in highbrow periodicals like *The New York Review of Books* and *The Poetry Review in Britain*; in general-interest publications like *Parade* and *USA Today*; in obscure foreign trade journals like *China Copyright* and *Svensk Bokhandel*; and in newspapers in places like Fiji, Greenland, the Falkland Islands, and the Polynesian island of Niue (the name is roughly translated as Behold the Coconut!), which has one newspaper.").

[44] See, for example, Roy Blount, "Let s Not Lose Our Heads Over a 'Monopoly' of Orphans," The Authors Guild, June 24, 2009, available at www.authorsguild.org/advocacy/articles/roy-blount-on-google-orphans.html.

[45] Grimmelmann, "The Google Book Search Settlement: Ends, Means, and the Future of Books,", supra note 3 at 8; Peter Hirtle, "Why the Google Books Settlement is Better Than Orphan Works Legislation," LawLibrary Blog, May 27, 2009, available at http://blog.librarylaw.com/librarylaw/2009/05/why-the-google-books-settlement-is-better-than-orphan-works-legislation.html.

[46] The minimum inclusion fee under the settlement is, for example, $60. Settlement Agreement, *Authors Guild, Inc. v.Google, Inc.*, Case No. CV 8136-JES § 2.1(b) [hereinafter Settlement Agreement].

[47] Blount, "Let's Not Lose Our Heads Over a 'Monopoly' of Orphans," supra note 44; see also Hirtle, "Why the Google Books Settlement is Better Than Orphan Works Legislation," supra note 45, (citing Carneige Mellon University study that located 79 percent of rightsholders).

[48] Hirtle, "Why the Google Books Settlement is Better Than Orphan Works Legislation," supra note 45.

[49] Kahle s criticisms of the settlement are unfounded. Kahle argued that the settlement should not be approved because Google would gain a monopoly over millions of orphan works yet also stated "[t]here are alternatives" to Google s scanning efforts and that entry at scale is not that expensive. Brewster Kahle, "A Book Grab by Google," *The Washington Post*, May 19, 2009, available at www.washingtonpost.com/wp-dyn/content /article/2009/05/18/AR2009051802637.html. Kahle does not appear to appreciate that monopolies do not arise where alternatives exist and entry is viable.

[50] Brewster Kahle, "Economics of Book Digitization," Open Content Alliance, available at http://www.opencontentalliance.org/2009/03/22/economics-of-book-digitization.

[51] Grimmelmann, "The Google Book Search Settlement: Ends, Means, and the Future of Books,", supra note 3 at 10.

[52] Ibid..

[53] Ibid..at 11.

[54] For example, Ibid..at 1 (Google developing a "dominant platform with control over a huge catalog of books that no one else has access to").

[55] Ibid at 5; Settlement Agreement, § 4.2.

[56] Besides the pricing options provided in the settlement, authors and publishers can also opt-out of the settlement and negotiate sales prices and other access provisions directly with Google. When seen in this broader context, it s even more difficult to contend that the consumer sale provisions will facilitate collusion.

[57] Settlement Agreement at § 4.2.

[58] *BMI*, 441 U.S. at 9 (further noting that "[l]iteralness is overly simplistic and often overbroad.").

[59] Ibid. at 18 (emphasis added).

[60] Ibid at 20.

[61] Grimmelmann, "The Google Book Search Settlement: Ends, Means, and the Future of Books,", supra note 3 at 6. He also noted that a "common theme" among his concerns "is that they all relate to, or are magnified by, centralized power." Ibid at 7.

[62] Ibid. at 14.

[63] Department of Justice Business Review Letter, "Copyright Clearance Center, Inc.," August 2, 1993, available at www.usdoj.gov/atr/public/busreview/211651.htm.

[64] Ibid.

[65] Ibid.

[66] Grimmelmann, "How to Fix the Google Book Search Settlement,", supra note 3 at 13.

[67] Settlement Agreement at § 6.1.

[68] Ibid at 6.

[69] See for example, David A. Balto, "Networks and Exclusivity: Antitrust Analysis to Promote Network Competition," *George Mason Law Review* 7 (523)(1999): 537-43.

[70] Settlement Agreement at § 3.8(a).

In: Digital Books: Competition and Commerce
Editor: Oliver A. Hagen
ISBN: 978-1-61728-299-7

Chapter 6

TESTIMONY OF DAVID DRUMMOND, SENIOR VICE PRESIDENT OF CORPORATE DEVELOPMENT AND CHIEF LEGAL OFFICER, GOOGLE INC., BEFORE THE HOUSE COMMITTEE ON THE JUDICIARY, HEARING ON "COMPETITION AND COMMERCE IN DIGITAL BOOKS"

David Drummond

Imagine if a student living in a rural area or inner city could go to her local public library and read from millions of books in the combined collections of some of our nation's greatest libraries, including the University of Michigan, the University of Texas at Austin, Stanford University, the University of Wisconsin-Madison, and the New York Public Library. Or if a blind student suddenly could access millions of digital books, using a refreshable Braille reader to unlock knowledge foreclosed from the visually impaired today. Then consider the author whose life's work, a book no longer in publication, suddenly becomes available online so that anyone could find it, buy it and read it.

That's why I am excited to be here today – because these are the types of opportunities that will be created by the settlement of a lawsuit brought against Google by authors and publishers. If approved by the U.S. District Court for the Southern District of New York, the settlement will create an educational, cultural and commercial platform to expand access to millions of books for anyone in the United States, enriching our country's cultural heritage and intellectual strength in the global economy.

As explained below, we believe Google Books is fully compliant with copyright law today, and, if approved, the settlement will expand access for the public while providing rightsholders choice and compensation; lower barriers for other entrants; complement orphan works legislative efforts; and preserve Congress's role in setting copyright policy.

But this settlement is not what is driving our digital books efforts. The world of online books is changing in far more revolutionary ways. From the Kindle to the cloud, companies are racing to offer new ways for readers to access digital books. Despite its cultural significance, our settlement is about the past – it only covers books published before January 5, 2009, most no longer in publication and with low commercial demand. It's the newest titles and books yet to be written that will drive competition and commerce for digital books, not out-of-print books held in libraries.

The Evolving Book Industry

The vast majority of books sold in the United States are in print. The $25 billion retail market skews heavily toward new releases, with out-of-print representing only two to three percent. Even used bookstores sell mostly in-print works. So don't let anyone fool you that the future of books depends on what happens with the settlement of a lawsuit over out-of-print library books.

What we anticipate will revolutionize the way some people read books is an open cloud-based platform, where users buy and store digital books in online personal libraries accessible from any Internet-connected device. Amazon's Kindle approach links its online bookstore with its hardware device in a proprietary system, where users buy their books and device from a single source – Amazon. We are partnering with bookstores, publishers, and device manufacturers to develop an open platform. In this open platform, readers can find and purchase digital books from any bookstore and read them on any device, including laptops, mobile phones, and e- readers from multiple vendors. Smaller, independent bookstores, such as BookPeople in Austin, will benefit from an open platform that helps them stay relevant as book consumption moves online. Ours and other retail syndication efforts may be critical for consumers to retain a diverse range of options when shopping for books.

Google does not currently sell books. We are a new entrant, starting with zero market share. We're excited that the settlement will let us sell out-of-print books online through an open platform to anyone in the United States. Indeed, we believe it may be the development of this open platform, rather than concern about the marginal economics of out-of-print works, that underlies corporate opposition to this agreement. After all, in a market driven overwhelmingly by new titles, sales from the settlement will be a tiny fraction of overall book sales. We didn't settle the lawsuit thinking it would catapult us to be the next Amazon – rather, opening up access to these books helps fulfill our founders' vision for our digitization efforts.

Google Books – Origin and Product Features

Larry Page and Sergey Brin, Google's founders, have long believed in the power of unlocking information contained in books. Their interest in making digital libraries accessible dates back to 1996 when they worked on the Stanford Digital Library Technology Project as graduate students.

When Larry and Sergey approached me with the bold vision of doing for books what we do for the web – namely, making copies and indexing the text to make it searchable – as the company's lawyer I was a little taken aback. But as I thought about it, it made perfect sense.

Today it is understood that the act of copying the web to index it is a fair use under our nation's copyright laws. Fair use is the very reason search engines exist. Well, the same fair use principles apply here – we make copies of books, index them, and only show users a few lines of text if the book is still under copyright protection. If you have used Google Books and seen more than short snippets, then you were looking at a book that was licensed to us or is in the public domain.

Since we started in 2004, we have scanned more than 10 million books. We get the books in two main ways. First, publishers partner with us on a non-exclusive basis to give us licenses to display and preview often their newest and commercially valuable titles. We have more than 30,000 publisher partners, representing more than two million books. For example, we partnered with the University of Michigan Press to make available a 2001 book entitled *Before Motown: A History of Jazz in Detroit 1920-1960*, for which users can browse several pages online. Our "Partner Program" was not the subject of the lawsuit or settlement.

The main way we get out-of-print books is from 42 global library partners, including 30 major institutions in the United States. Libraries have long been entrusted to preserve information spanning the range of human knowledge. Our partners have enriching historic collections, such as the Nettie Lee Benson Collection at the University of Texas comprising rare books that chronicle the history, politics and society of Latin America. Our library agreements are nonexclusive, and several participating libraries partner with other digitization efforts.

Of the 10 million books, we estimate that at least two million are clearly in the public domain and not part of the lawsuit. These include books published before 1923. For these books, when a user enters a relevant query we display the entire text. Users can freely download the book in multiple formats. Making public domain texts discoverable online is already advancing education and scholarly pursuits. Tim Barton of Oxford University Press describes a Columbia University classics class assignment, in which 70 percent of the undergraduates cited a book published in 1900 that was not on the reading list and long overlooked in classics scholarship.

That leaves six million books that are likely still under copyright protection but mostly out-of-print, although not all orphaned. For these books, we currently show users three small snippets of text around the search term, often just a few lines from the book. This snippet view is not particularly useful to our users – they cannot replicate the experience in physical bookstores, where people tend to pick up books and browse through a few pages before buying. Google Books also provides links to show users where to locate the book in a library or buy it from retailers. We drive traffic today for free to Amazon, Barnes & Noble, used book stores, and other booksellers through these links.

Even with our limited uses of library scans and restricted snippet view, in 2005 authors and publishers sued us in separate lawsuits, with the authors filing a class action. The lawsuits essentially claimed copyright violations for scanning, indexing and displaying portions of in-copyright works. We strongly disagreed then, and we still disagree today. Nothing about the settlement changes our firm belief that copying for the purpose of indexing is a fair use that is encouraged by existing copyright law precedents. Fair use is critical to the way web search and book search work, and Google vigorously defends fair use in this and other contexts.

SETTLEMENT EXPANDS ACCESS TO LOCKED-UP INFORMATION AND BENEFITS COPYRIGHT OWNERS

As the lawsuit progressed, at the authors' request, we sought to find common ground among authors, publishers, libraries and Google to expand distribution channels and make out-of-print books more accessible. In October 2008, we announced a settlement that will:

- Allow anyone, anywhere in the U.S. to preview out-of-print books and purchase an online version, right from their computer;
- Provide every public and university library across the U.S. free viewing of millions of books at a designated computer, plus the ability to purchase an institutional subscription;
- Provide the print disabled and visually impaired unprecedented access to the written literary record;
- Create new opportunities for authors and publishers to sell their books; and
- Enhance innovation and competition in digital books by creating a non-exclusive registry to clear rights, collect revenue and license works efficiently to Google and others.

The Reading Public – Google settled the case mainly so we can provide readers greater access to books. Rather than continue showing only snippets of out-of-print books, the settlement will allow us to display previews of up to 20 percent of the book, sell digital versions and provide access to institutional subscriptions, unless the rightsholder instructs us otherwise. With strong privacy protections, users will be able to browse and buy digital copies of millions of books that otherwise might be left behind in the digital age.

Out-of-print books are not sold through most bookstores and typically are found only in a limited number of research libraries, making access difficult or time consuming for much of the population. Expanded access will broadly benefit readers, researchers, and students, but it will be felt most tangibly by those who historically have had the least access to books, particularly those living in under-funded areas or with a disability that hinders traveling to, or reading from, their local libraries. Regardless of geography, income or physical disabilities, the settlement will greatly expand access to the world of knowledge contained in our nation's largest libraries. According to Wade Henderson of the Leadership Conference on Civil Rights, "[f]ailure to approve the settlement would be tragic, in that it would impede meaningful access to vital information for many of who have been denied for far too long."

Libraries and Academic Institutions – The settlement will let Google sell an institutional subscription enabling libraries, universities and other organizations to offer access to millions of books from the world's leading libraries. Expanded access to these vast collections will serve as an important equalizer. Historically black colleges and community colleges are eager to attract faculty and students and level the playing field with larger institutions. The institutional subscription will be priced to assure a market rate for the rightsholder and broad access by the public. Under the settlement, public libraries and non-profit higher education institutions can obtain free access to the institutional database at one on-site computer.

Scholars and Researchers – Many scholars and librarians support our work as complementary to their efforts to make books accessible in an increasingly digital society. Gregory Crane, who runs the Perseus Digital Library Project to preserve Greco-Roman literature and culture, believes the settlement "is a watershed event and can serve as a catalyst for the reinvention of education, research and intellectual life." Google also will provide $5 million to create two research centers for computational research across the corpus.

Disability Access – Millions of print-disabled Americans will receive revolutionary new access to books through the settlement. As CNET reported earlier this month:

> "Blind people ... have access to a special library run by the Library of Congress that converts print books into formats readable by the visually impaired, but that library – in existence since 1931 – only has 70,000 texts, said Chris Danielsen, director of public relations for the National Federation of the Blind. If the settlement is approved in October, it will give 'print-disabled' people 'access to more books than we have ever had in human history,' he said."

After the settlement, millions of out-of-print texts will be accessible via screen enlargement, screen reader, and refreshable Braille display technologies. The National Federation of the Blind believes this will be pivotal in shifting the current inaccessible e-book archetype to one that assures equal access. For the first time, the print disabled will access our printed heritage to a degree comparable to that enjoyed by other Americans.

Authors and Publishers – The settlement provides a means to locate and compensate authors whose works might otherwise never be distributed online, while preserving rightsholder choice. Under the settlement, the rightsholder is in control, regardless of whether the book is in print or out of print. At any time, the rightsholder can direct Google to turn displays on or off, start or stop selling digital versions, or not scan particular books. Rightsholders will receive 63 percent of revenue earned from purchases, advertising, and subscription sales. For consumer purchases, rightsholders can set their own price (including a price of zero, as somc authors simply want to give away their older books), or they can choose to have Google set a competitive price using an algorithm. The registry will provide authors a low-cost mechanism to resolve ambiguity over digital rights and license works to other providers in addition to Google.

Settlement Lowers Barriers for Other Entrants

We are the only company to date that has attempted to digitize the vast in-copyright collections of U.S. libraries. Even so, nothing in the settlement prevents anyone from doing what we have done. The agreement is non-exclusive in every possible respect, and the creation of the registry will make it easier for other companies to enter the market. That is why several leading antitrust scholars have praised the settlement's pro-competitive benefits. Indeed, e-reader manufacturer Sony Electronics believes the settlement "will foster competition, spur innovation, and create efficiencies that will substantially benefit consumers."

Search Engines – Any search engine that wants to scan and index in-copyright books can already do so. It is understood that scanning for the purpose of indexing is fair use. The settlement therefore has no affect on the ability of other search engines to compete by scanning and indexing in-copyright books, whether orphaned or not. Microsoft announced its own digitization initiative in 2005 for public domain and partner books. Three years later, after scanning 750,000 books, Microsoft shut down the program for financial reasons, preferring to crawl the repositories created by others instead. After the settlement, just as before, search engines no doubt will continue making their own business decisions about digitization.

Book Retailers – Next is the question of whether the settlement harms competition for book sales. Because the settlement largely deals with out-of-print works, it does little to change the state of competition for new, in-print titles, which comprise 97 percent of the market. Through Google Editions, we will work with publishers to offer digital versions of in-print books through multiple retailers. Our open retail platform will expand distribution in a way that is clearly beneficial for the public.

Competition for out-of-print books is restrained for everyone. Unlike newer titles, as a practical matter it is nearly impossible to clear ownership rights to millions of older, out-of-print books. It's not simply a matter of locating the author. A tangle of legal uncertainties apply, including decades-old contracts, different copyright regimes depending on when and where in the world the book was published, whether the book was registered or renewed with the Copyright Office, and whether the inserts, illustrations or images have separate copyright holders. The right to publish the hard copy may or may not include digital rights, an unresolved legal entanglement between authors and publishers. Long ago communications between author and publisher may have left either one with the sole ability to license the work. Facing the threat of statutory damages liability (as much as $30,000 per work) for making a mistake even if the rightsholder is not harmed, often neither author nor publisher is willing to commit that they can license the work for online use. The cost of ascertaining for certain who has the rights to an out-of-print book likely exceeds the economic potential of any given book. The result is that these books, which collectively represent much of our nation's printed cultural heritage, have remained inaccessible.

Nothing in the settlement makes it any more difficult for others to license these books. Rather, the settlement is structured to make it easier for anyone – including Google's competitors – to clear rights. Google is funding a non-profit registry, controlled by authors and publishers, to resolve ownership disputes. As rightsholders come forward, information about what books were claimed and who claimed them will be publicly available, thereby lowering the costs and risks for other providers. Google's competitors can use this information to avoid scanning books with limited market potential and focus on commercially valuable works. Later entrants also can take advantage of Google's efforts to identify books in the public domain. Moreover, while rightsholders retain licensing of their own works, they also can authorize the registry to license to third parties to the extent allowed by law. Many books that were once difficult for anyone to license will become books that are easy for everyone to license.

If within ten years the registry licenses a significant number of unclaimed works to another distributor on more favorable terms (less favorable for the rightsholder), Google can receive equal treatment. This simply protects against others free riding off the investment of

Google in creating the registry, and reflects the fact that, unlike Google whose terms are fixed through the settlement, competitors can negotiate terms based on future market realities. Moreover, the provision does not apply to claimed books, which hold the lion's share of economic value. Importantly, it does nothing to reduce the registry's incentives to license our competitors. If a competitor offers the registry a better deal, the registry has every incentive to take it, and the provision doesn't apply. While this clause has generated much controversy, the structure is relatively commonplace and understood to be pro-competitive in contexts like this.

Settlement Reduces the Problem of Orphan Works for Everyone

Many people have expressed concern about the problem of "orphan works," a concern we share. We have long supported an effective orphan works legislative solution, and we will continue to do so. This settlement is a strong complement to, and not a substitute for, orphan works legislation.

An "orphaned" book is an abandoned book. Many out-of-print books, however, are not abandoned but instead have two parents whose ownership rights are uncertain. We call these books "neglected" books. Data points indicate that orphaned books are at most about 20 percent of out-of-print books, and likely would be lower with a financial incentive and efficient mechanism for rightsholders to claim books and clear rights, which the settlement provides. Even among the orphans, a substantial portion may be commercially insignificant or miscellaneous works, such as a transistor handbook from 1966 that we scanned and is now likely abandoned.

Over time, the structure of the settlement will reduce the orphan work problem for everyone. The settlement will create a registry whose job it is to go out and find rightsholders. It also will create a financial incentive for rightsholders to come forward. And it creates a database that identifies these rightsholders, making it easy for other providers to find them and obtain a license for their books. We believe over time a significant number of works will be claimed. Ideally, the registry would eliminate orphan books altogether. But realistically some small portion of rightsholders will still be unable to be found, resulting in a true abandoned book, or orphan.

The settlement will enable Google to make certain uses of abandoned books. So far we are the only company that has sought to digitize in-copyright, potentially orphaned books. We believe anyone who wants to re-use abandoned works should have a fair, legal way to do so. In our view, the settlement helps here too.

To the extent that other providers want a legal framework to re-use orphaned books (without having to defend against a lawsuit like we did), Google would support a legislative solution. The settlement provides a working model – a private sector incentive and mechanism to clear rights – that will reveal the scope of the orphan problem and spur legislation. Indeed, past legislative efforts have contemplated and encouraged private sector initiatives to build rights databases like the one Google will fund, which the Copyright Office called "indispensable" to solving the orphan works problem.

Some scholars have taken a well-intentioned view that it is preferable to pass orphan works legislation before letting Google make abandoned books more accessible. This would mean we all wait for enactment of a broader bill that addresses not only orphan works but also rights clearance challenges for neglected works. In the meantime, instead of having multiple potential providers of neglected books and at least one provider of abandoned books, we will have no providers of any of these books. Moreover, absent the settlement or a government-funded registry, it remains unclear whether the private sector will fund the necessary databases.

For truly orphaned books, why is one provider now better than none? Because one company spending the resources to make orphaned books more accessible greatly enriches our cultural heritage and expands the progress of human knowledge. Every day, older works in libraries are being taken off shelves and sent to storage facilities. Given the opportunity to revive access to these books for the public in a way that harms neither competition nor commerce, nor the chance of enacting future legislation, I would hope the choice is obvious.

THE SETTLEMENT IN CONTEXT – CONGRESS CONTINUES ITS LEGISLATIVE ROLE

The settlement represents the resolution of a long and hard-fought litigation among multiple parties with divergent interests. The suggestion that the settlement usurps the role of Congress to set copyright policy because the suit took the form of a class action is flatly wrong. The settlement does not establish new copyright law; it is not even a determination on the merits of copyright law. All the settlement represents is the means by which the class of rightsholders decided to resolve the lawsuit.

Critics may dislike the use of class actions in copyright, but Congress itself created class actions through Rule 23 of the Federal Rules of Civil Procedure and has not restricted their use in copyright cases. Copyright class actions are not uncommon, and have been settled through similar remedies. The class action process allows rightsholders to object to the settlement or opt out. Indeed, this settlement will let rightsholders control the use of their books even after the settlement takes effect. Moreover, as the means of redress for private litigation in the United States, the settlement is fully consistent with international treaty obligations – a view confirmed by Berne Convention scholar Sam Ricketson.

I've heard many suggestions for improving the settlement, some diametrically opposed. But the Judge's role is to approve or reject. If rejected, the parties likely return to their litigation stances, arguing over snippets and indexing, losing this opportunity to open up online access to information trapped in out-of-print books.

It would be unfortunate if this hearing devolved into hypothetical debates over class action law in the copyright context, postulation of the Platonic ideal of orphan works legislation, or simply a forum for competitors to argue over what is estimated to be less than three percent of the commercial market for books. Something far greater for human knowledge is at stake.

The very purpose of Article I, Section 8, Clause 8, the copyright clause, of the Constitution is to "promote the Progress of Science and useful Arts." As reading moves online, and new generations of students tap into centuries of learning from their laptops, users will find knowledge they greatly value but did not even know existed. While this may not generate significant commercial value, it will elevate the marketplace of ideas. The settlement represents the progress of science to tackle copyright challenges and help ensure millions of out-of-print books do not fade into oblivion. To oppose this settlement means depriving the public of learning, and punishing the parties to a lawsuit for resolving their private litigation in innovative and groundbreaking ways.

In: Digital Books: Competition and Commerce
Editor: Oliver A. Hagen
ISBN: 978-1-61728-299-7

Chapter 7

STATEMENT OF PAUL MISENER, VICE PRESIDENT, GLOBAL PUBLIC POLICY, AMAZON.COM, BEFORE THE COMMITTEE ON THE JUDICIARY, HEARING ON "COMPETITION AND COMMERCE IN DIGITAL BOOKS"

Paul Misener

Good morning, Chairman Conyers, Ranking Member Smith, and members of the Committee. My name is Paul Misener, and I am Amazon.com's Vice President for Global Public Policy. Thank you for inviting me to testify on this important topic.

On behalf of our customers and company, and in the interests of promoting competition and commerce in digital books, Amazon.com strongly objects to the proposed class action settlement among Google, the Authors Guild, and some book publishers. If approved, this proposal would create national copyright and competition policy with enduring adverse effects on consumers and Google's competitors.

Amazon has joined librarians, legal scholars, authors, publishers, and other technology companies in the Open Book Alliance to counter the proposed settlement.[*]

Our shared view is that we strongly support mass digitization and electronic distribution of books, but we insist that any such efforts be undertaken in the open, grounded in sound public policy, and mindful of the need to promote long-term benefits for consumers rather than those of a few commercial interests.

Mr. Chairman, Amazon takes no pleasure in opposing Google in the class action case or in today's hearing. As you may recall, we work closely with Google on other matters before your Committee, including net neutrality, where we both want rules to protect consumers in the absence of competition, and Google has often said they seek such rules to enable "the next Google."

---

* The Alliance's membership includes the Special Libraries Association, the New York Library Association, the American Society of Journalists and Authors, the Council of Literary Magazines and Presses, Small Press Distribution, the Internet Archive, Microsoft, and Yahoo! *See* www.openbookalliance.org.

Similarly for book scanning and distribution, Amazon seeks to preserve the competitive market so that we may continue to provide our customers the great selection they have come to expect from us. Unfortunately, however, under the threat of liability for what the Authors Guild called "massive copyright infringement," Google has taken another course. Rather than ask Congress to protect consumers and competition, Google instead has asked a trial court to approve a class action settlement that would establish national copyright and competition policy exclusively in favor of Google above all potential competitors.

Mr. Chairman, the proposed settlement is exceedingly complex, primarily because it is much more of a joint venture agreement and establishment of national policy, than a resolution of claims arising from past behavior. The proposal contains many flaws that run counter to consumers' interests but, in light of our limited time today, let me focus on two flaws, either of which is sufficient to condemn and reject the proposed settlement.

First, the proposal would create a cartel of rightsholders that, for sales of books to consumers, would set prices to maximize revenues to cartel members. This cartel, called the Book Rights Registry, could never have been established in the ordinary course of business. Currently, rightsholders may individually license their works for electronic distribution and, thus, compete against each other. But the proposed Registry is based on an agreement among erstwhile competitors to collectively set prices for their products. The Registry is a classic horizontal cartel and, with respect to electronic books sold to consumers (including those ultimately printed on paper using "print on demand" technology), the Registry cartel is novel mostly for the automation it applies to price setting. Instead of a cigar smoke-filled room, the Registry cartel would set book prices within the clean confines of a computer "Pricing Algorithm" designed by Google to "find the optimal such price for each Book and, accordingly, to maximize revenue for each Rightsholder." This algorithm could optimize the price for a certain set of books simply by increasing all prices of similar books at the same time – a nifty, smokeless feat of collusion for revenue maximization.

Making matters much worse, in the proposed settlement's second fatal flaw, Google would get a privileged, exclusive deal, despite lip service to non-exclusivity. Except for works of rightsholders who affirmatively opt out, the settlement would give Google – and only Google – a license to digitize and sell every U.S. book ever written. This means that Google alone would have a permanent and exclusive right to copy, display, or sell digital versions of the millions of orphan works. It is nonsense to claim that potential Google competitors would have access to the same deal, either directly, through the Registry, or by following Google's courtroom odyssey. For one thing, the Registry cannot license competitors to scan orphan works because it can only license uses of books whose copyright owners have given their express approval. Moreover, it's not at all clear whether the Registry would be willing to license these registered works to compete with the Google deal. And, if a potential competitor to Google engaged in "massive copyright infringement" in the hopes of getting sued by the same plaintiffs in the Google litigation and making the same settlement deal, why would the rightsholders settle on the same terms when they already have a distribution partner and would stand a reasonable chance of obtaining massive statutory damages? In any case, the "MFN clause" of the proposed settlement would guarantee that a hopeful competitor to Google could not get a better deal.

As a result of these flaws, the proposed settlement would seriously harm individual book consumers and most libraries and schools because the rightsholders cartel and Google monopoly inevitably would set higher prices or provide worse service than would be available in a competitive market. The proposal also would harm existing and potential Google competitors, who would be denied a fair and reasonable opportunity to license a similar corpus of works under similar terms. Under the proposed settlement, "the next Google" wouldn't stand a chance, and customers of existing Google competitors would, instead of realizing the myriad benefits of market choices, find themselves at the mercy of a sole source provider. Under the proposed settlement, Google would become a consumer's nightmare: the only store in town.

Amazingly, the proposal also would exonerate Google of future claims based on future actions that would otherwise be prohibited by law. This is an impermissible result of class action litigation and, again, makes the proposed settlement less about resolution of a legal dispute than about copyright policymaking and forming a joint venture.

Mr. Chairman, if you have not already, you undoubtedly will hear from Google and some others about the potential consumer benefits of the proposed settlement. Amazon has its own book scanning project, and clearly recognizes the benefits of digitization and distribution, and would welcome a statutory solution to the orphan works problem. But even if this matter were being evaluated purely on policy grounds, the costs of the proposed settlement far outweigh the potential benefits. A price-setting cartel and monopoly are means emphatically *not* justified by the ends, especially because legislation could produce the same benefits in a pro-consumer, pro-competitive manner.

Indeed, the novel copyright (and competition and class action litigation) policy matters at issue in the proposed settlement should be addressed in Congress, the appropriate venue for national policymaking. We hope that the court will not approve the proposed settlement when it acts, as expected, in the coming weeks. In the interim, we merely ask that this Committee carefully monitor developments in the case. If the court approves the proposal without fixing its serious anti-consumer, anti-competitive flaws, we respectfully request that Congress reestablish its public policy making authority and act quickly to supersede the settlement with appropriate legislation.

Thank you again for the opportunity to testify. I look forward to your questions.

In: Digital Books: Competition and Commerce
Editor: Oliver A. Hagen
ISBN: 978-1-61728-299-7

Chapter 8

WRITTEN TESTIMONY OF RANDAL C. PICKER, PAUL AND THE LEFFMANN PROFESSOR OF COMMERICAL LAW, THE UNIVERSITY OF CHICAGO LAW SCHOOL AND SENIOR FELLOW, THE COMPUTATION INSTITUTE, THE UNIVERSITY OF CHICAGO AND ARGONNE NATIONAL LABORATORY, HEARING ON "COMPETITION AND COMMERCE IN DIGITAL BOOKS"

Randal C. Picker

Mr. Chairman and Members of the Committee:

Thank you for the invitation to testify today on competition and commerce issues in digital books. My name is Randal C. Picker. I am the Paul and Theo Leffmann Professor of Commercial Law at the University of Chicago Law School. I am also a Senior Fellow at the Computation Institute of The University of Chicago and Argonne National Laboratory. I have taught at Chicago since 1989 and write and teach in a number of areas including, of relevance to today's hearing, antitrust, copyright and network industries.

Faculty offices at the University of Chicago Law School are physically in the D'Angelo Law Library. I walk out of my office door into book stacks and I feel very lucky to have ready access to one of the world's great university libraries. Yet notwithstanding that, I have an almost unnatural level of affection for Google Book Search. If you haven't used it, you should, as it is a wonder. I have been doing research into some business practices in the early 1900s. Google Book Search is a powerful window into the past and one that I can look through from any computer anywhere. And, again, I say that as someone who has a great library literally outside his office door. Imagine what that access means for people who are less fortunate.

But creating a great new product doesn't somehow entitle a firm to a broad exemption from the law of the land. We wouldn't consider for an instant allowing Google or any other firm to violate environmental laws or civil rights laws merely because they were creating a great new product. You have to do both: innovate and comply with the law. More precisely here, the fact that Google Book Search generates substantial benefits to consumers does not somehow insulate the Google Book Search settlement from antitrust inquiry. We should expect antitrust officials to assess whether those benefits can be preserved while at the same time minimizing any possible anticompetitive features of the settlement. Antitrust officials do not and should not allow firms to engage in anti-competitive practices merely because those practices have been bundled into a larger project that has substantial competitive benefits. Antitrust regulators need to separate the pro-competitive wheat from the anti-competitive chaff.

Earlier this year, I was asked to speak at a conference at Columbia Law school on antitrust and competition issues in the GBS settlement. In connection with preparing for that talk, I commenced a draft of the paper on those issues and subsequently completed it after the conference. That article was recently published in the September, 2009 issue of the *Journal of Competition, Law and Economics*. A preprint version of that article is attached to this testimony and my testimony is largely based on the article, though my thinking about the consequences of the GBS settlement has continued to evolve and there are some new thoughts in this testimony.

I make three central points in the article and those points are addressed to three different government audiences, first to the courts, then to the Antitrust Division in the Department of Justice, and then to the courts but really to Congress. Take those points one by one.

1. *No* Antitrust *Immunity from Approval of the Settlement*. One point is a particularly technical issue of antitrust law, namely whether court approval of the GBS settlement will insulate it from subsequent antitrust inquiry. This is a question of the scope of the *Noerr-Pennington* doctrine of antitrust immunity. That point is addressed to the court system and first to Judge Chin in particular who has the proposed settlement before him. Antitrust law provides that parties can petition the government and ask for competitive benefits without fear of antitrust prosecution, but the precise boundaries of that doctrine are uncertain. Should Judge Chin approve the settlement agreement, I don't believe that *Noerr-Pennington* immunity should attach, but I don't think the cases on this are completely clear. Should he approve the agreement, I think Judge Chin should add a provision stating that it is not his intent for *Noerr-Pennington* immunity to attach. I call this a no-*Noerr* clause. I should say that I made this up out of whole cloth and have no idea whether a subsequent court would respect that intent.

2. *Collective Pricing of Consumer Purchases as Sherman Act Section 1 Violation*. The second point is about the pricing mechanism for consumer purchases in Google Book Search. That point is addressed to our antitrust prosecutors. The Department of Justice is expected to file its views in this case by September 18, 2009. There are two core business models contemplated by the settlement. One is a blanket-license subscription for institutions to confer online bulk access to all of the works in Google Book Search. Although there have been concerns raised about the pricing of these

blanket licenses, my paper says very little about those. Under current United States antitrust doctrine, high prices associated with legitimately obtain monopolies do not give rise to antitrust violations. We accept those prices as a consequence of successful competition. My paper instead addresses the second business of the settlement, namely sales of individual access to online books to consumers. This is the online equivalent of going into Borders to buy a single copy of a book.
The settlement agreement implements a pricing algorithm for these individual consumer sales that is, like almost everything else in the agreement, quite complex. That said, I suggest in my article that there are ways that the algorithm can result in coordinated pricing among individual authors in a way that could give rise to problems under Section 1 of the Sherman Act which forbids much joint pricing as collusive. I don't regard that as a slam dunk argument, as I think that the pricing algorithm raises new issues, but I think that there is a good chance that the Department of Justice will address it in its filing.

3. *The Government Should Not Issue Just One License for the Use of Orphan Works.* The third point is addressed to Congress first and foremost but also to the court system. A key feature of the class-action settlement is that it will make available to Google — and only Google — the so-called orphan works. These are works that remain in copyright but where the holder of the copyright cannot be tracked down so that his or her work could be used with permission. I have seen figures from Google suggesting that perhaps as many as 80% of the works in GBS are orphans. The opt-out class action inverts the usual rule of copyright on its head. Usually someone like Google would need to get permission to use a copyrighted work unless its use would qualify as a fair use. Absent fair use, you have to ask to use a copyrighted work, and the problem with orphans is that you don't know who to ask.

The opt-out class action changes this. Orphan rightsholders need to come forward to be excluded from the settlement, and if they don't, Google will get full rights to use those works. Usually if you ask a copyright holder for permission and you don't get a response, you get nothing. You get no right to use the work at all. But here with the opt- out class action, silence is indeed golden: Google will get full rights to use the orphan works.

Google and the Authors Guild could implement large chunks of their agreement as a private deal without court approval. The key point of running this business plan through the court system is that it will give Google a government license to use the orphan works. No private party can create a right to use the orphan works. Only the government can do so either through some form of opt-out class action or if Congress enacts broader orphan-works legislation.

That leads to a number of natural questions. Under what circumstances should the government create a right to use the orphan works? Should the government create a right in favor of only one firm and thereby create a monopoly over those works or should the government grant multiple licenses to spur competition? If Google or Microsoft or Amazon came to Congress seeking orphan-works legislation, could we imagine new copyright legislation that granted a right in favor of only one of those companies and not the others? Given the role of the government in authorizing the use of the orphan works, the government will choose precisely how much competition is possible. If only one firm is authorized to use those works, we will have a monopoly over those works.

My paper discusses ways in which the federal district court considering the settlement agreement might consider broadening that agreement to allow other firms to use the orphan works. But there is a more direct route available and that would be for Congress to pass orphan-works legislation. In considering that possibility, it is important to recognize how different the proposed settlement is from orphan-works legislation has been proposed in the past. That legislation typically has contemplated that an extensive search would be performed by the person or firm contemplating the use of the work that it believes to be an orphan. The only way to know that you have an orphan is to look for the owner.

The problem with that is that it imposes a substantial burden—indeed perhaps a crippling burden—on mass digitization efforts of the sort seen in Google Book Search. Anything that would require human beings to search individually as opposed to computers proceeding automatically would, I suspect, raise costs sufficiently to block most large-scale digitization efforts. In contrast, the Google Book Search settlement requires no such search in advance to include orphan works in the database. Moving from a required-search version of orphan-works legislation to something closer to the Google Book Search settlement might raise issues under the Berne convention, but those might be sidestepped by limiting the orphan-works legislation to United States works akin to what we do for the registration rules set out in Section 411 of the Copyright Act.

To close, we might return to the early 1900s and the days of player pianos. It wasn't clear whether the piano rolls infringed on the underlying musical composition that they played. In 1908, in the *White-Smith* case, the Supreme Court ruled that the piano rolls weren't copies, but Congress planned to overturn that result in new legislation. Knowing that, Aeolian, the leading producer of player pianos, entered into widespread contracts to lock up access to the new player piano roll rights. Faced with the threat of a monopoly over those rights, Congress created the first compulsory license, the so-called mechanical license. Under that license, once a composer had licensed his or her work to one firm, a second firm could pay a statutory fee and making competing piano rolls.

Only the government can license use of the orphan works and the government will choose how many licenses to grant. One license means monopoly, open-licensing the possibility of meaningful competition. Congress should pass orphan-works legislation. To its great credit, Google remains in favor of such legislation. That legislation should track many of the features of the Google Book Search settlement so as to enable full competition in the use of orphan works. Only the government can make that competition possible.

THE GOOGLE BOOK SEARCH SETTLEMENT: A NEW ORPHAN-WORKS MONOPOLY?

Randal C. Picker*

ABSTRACT

This paper considers the proposed settlement agreement between Google and the Authors Guild relating to Google Book Search ("GBS"). I focus on three issues that raise antitrust and competition policy concerns. First, the agreement calls for Google to act as agent for rights holders in setting the price of online access to consumers. Google is tasked with developing a pricing algorithm that will maximize revenues for each of those works. Direct competition among rights holders would push prices towards some measure of costs and would not be designed to maximize revenues. The consumer access pricing provision might very well fail a challenge under Section 1 of the Sherman Act. Second, and much more centrally to the settlement agreement, the opt-out class action will make it possible for Google to include orphan works in its book search service. Orphan works are works as to which the rightsholder cannot be identified or found. The opt-out class action is the vehicle for large-scale collective action by active rights holders. Active rights holders have little incentive to compete with themselves by granting multiple licenses of their works or of the orphan works. Plus under the terms of the settlement agreement, active rights holders benefit directly from the revenues attributable to orphan works used in GBS. We can mitigate the market power that will otherwise arise through the settlement by expanding the number of rights licenses available under the settlement agreement. To do that, we should take the step of unbundling the orphan works deal from the overall settlement agreement and create a separate license to use those works. All of that will undoubtedly add more complexity to what is already a large piece of work, and it may make sense to push out the new licenses to the future. That would mean ensuring now that the court retains jurisdiction to do that and/or giving the new registry created in the settlement the power to do this sort of licensing. Third, there is a risk that approval by the court of the settlement could cause antitrust immunities to attach to the arrangements created by the settlement agreement. As it is highly unlikely that the fairness hearing will undertake a meaningful antitrust analysis of those arrangements, if the district court approves the settlement, the court should include a clause—call this a no *Noerr* clause—in the order approving the settlement providing that no antitrust immunities attach from the court's approval.

JEL: D4; K20; K21; K41; L4; L43; O34

 Paul and Theo Leffmann Professor of Commercial Law, The University of Chicago Law School and Senior Fellow, The Computation Institute of the University of Chicago and Argonne National Laboratory. This paper is based on a talk that I gave on March 13, 2009 at the conference "The Google Books Settlement: What Will It Mean for the Long Term" at Columbia Law School's Kernochan Center for Law, Media and the Arts. I thank the Kernochan Center for hosting the conference and June Besek in particular for inviting me. I thank the John M. Olin Foundation and the Paul H. Leffmann Fund for their generous research support. I also thank Matthew Stoker for research assistance. Three disclaimers. First, I have served as a consultant in connection with the opposition to Google's proposed deals with Double Click and Yahoo!. I think that nothing there bears directly on the issues raised in this paper. Second, I have received research grants from Microsoft directly and the Law School receives funding from Microsoft. Again, I think that none of these bear directly on this paper. Third, and finally, I currently chair the faculty board for the Library at the University of Chicago. Our library is a member of the Committee on Institutional Cooperation, a library consortium of Chicago and the eleven schools in the Big Ten. As a group, the CIC has been library partners with Google since June, 2007.

I. Introduction

Google is a company of modest ambitions. As it says in its brief corporate statement, Google's mission is to "organize the world's information and make it universally accessible and useful."[1] Organize it, put it online, display it and make a few dollars at the same time. Google's Book Search is a core piece of this vision. Think of the world's great libraries, all merged into one collection and all available online through any device connected to the Internet. Universal access indeed.

But creating such a wonder is not a simple undertaking. Books have to be found, bought or borrowed and copied. The resulting digital files need to be sliced and diced to make them as useful as possible but also preserved so that looking at books online is very much like looking at them offline. This is a substantial technical undertaking, plus we need to figure out a business model for accessing the books. In the past—and still today of course—individuals purchased books or borrowed them from libraries, who in turn had purchased the books. Will digital copies be purchased in the same fashion or will different rules apply? Were these technical and economic challenges not enough, we would confront the really hard problem, namely, how do we match an 18th century legal system with early 21st century opportunities?

Google moved forward nonetheless. That in turn led to two lawsuits and an eventual settlement agreement that will be considered at a fairness hearing in federal district court on October 7, 2009. The settlement agreement is exceedingly complex—though not obviously unnecessarily so—as befits an agreement that will create an extraordinary new platform for accessing books.[2] Successful new book platforms are rare—since Gutenberg have there been any?—and Google's is of breathtaking ambition.

This paper considers some of the antitrust and competition policy issues raised by the settlement agreement. The paper itself is divided into seven sections. Section II provides brief background to the creation of Google Book Search and the lawsuits that emerged. Section III sets out five quick situations— hypotheticals, as we call them in law school—to try to establish some antitrust bearings to help us triangulate on the settlement agreement.

Section IV sets out some of the salient features of the settlement agreement. Absent the lawsuit by the Authors Guild, the settlement agreement would be nothing more than a private contract between Google and individual rightsholders with both horizontal and vertical components. The lawsuit does not change that essentially though it does have the key consequence of bringing so-called orphan works within the agreement. These are works that remain within copyright but that are stuck in limbo: the rightsholder for the book cannot be identified or, if identifiable, cannot be tracked down. That means that it is not possible to license access to the work. You cannot contract with people you cannot identify or find.

That takes us to two antitrust/competition policy issues and then to a key question of timing and comparative institutional advantage. First, the settlement agreement implements a pricing algorithm for single-copy access to digital books that I think is questionable. This is a joint agreement among rightsholders with Google as to how Google will price online access to their works going forward. Rightsholders can choose to appoint Google as their agent for pricing online access to consumers where Google will seek to maximize revenues for each book. That is not the result that would emerge under pure competition between authors/rightsholders and seems likely to run afoul of Section 1 of the Sherman Act.

Second, as currently configured, the settlement agreement creates unique access for Google to orphan works. This emerges directly from the court's presence. Absent the lawsuit by the Authors Guild, Google and interested rightsholders could have crafted a deal very much like that in the settlement agreement and would have implemented that through private contracts. That deal, of course, would be subject to antitrust scrutiny, as it would involve large numbers of otherwise competing rightsholders contracting together with Google. That would not be unprecedented—we have similarly complex arrangements for other copyright collectives like ASCAP and BMI—but definitely worth antitrust attention (70 years worth for our music cooperatives).

But with the lawsuit and the opt-out class action, we have left the world of purely-private contracts. For some rightsholders, that change is just a bother: they would not have had to sign a private deal and could have done nothing but now must affirmatively opt out of the settlement. But for our orphan rightsholders, the change in default positions is everything. The orphan holders cannot act and the settlement agreement neatly sweeps them up to give Google releases for the ongoing use of their works. The settlement agreement solves this problem for Google, but only for Google, in creating a carefully-crafted license for Google to use those works. The great accomplishment of the settlement is precisely in the way that it uses the opt-out class action to sidestep the orphan works problem. But, as has been noted elsewhere,[3] this gives Google an initial monopoly—and possibly a long-running one—over the use of the orphan works. This emerges *directly* from the court's role in this case because, again, the settlement agreement between Google and active rightsholders could have been implemented as a private matter without a lawsuit, though, again, with perhaps substantial antitrust attention. But the lawsuit is the device by which the initial orphan works monopoly is created: without the lawsuit, Google would acquire *no* rights to use the orphan works.

The way to prevent the creation of the market power that will arise directly from the court's role in approving the settlement would be to modify the settlement agreement by expanding the licenses created under it. I consider this issue in Section V. I think that there are strong reasons to think that the license created by the settlement agreement should be expanded so as to mitigate the market power that the court's approval of the settlement agreement will otherwise create. The most natural hook for this substantively would be the agreement's most-favored nations clause, which currently runs only in favor of Google. A more symmetric MFN would create a going-forward licensing mechanism for other entities to license the works of the active rightsholders as well to use the orphan works.

But I do think that there is a timing issue on that. Without real parties before the court on this, we are just shadowboxing. I do not know that I would modify the MFN clause in the abstract; we should probably wait instead until we have an actual case before us. The settlement agreement provides that the court will retain jurisdiction over it going forward. That jurisdiction needs to include the possibility that other parties can subsequently come to court and seek licenses. Another possibility is to ensure that the registry created under the settlement agreement has the power to issue licenses going forward. And there is a plausible reading of the settlement agreement that suggests that the registry is intended to have the authority to license the orphan works to third parties.

There is a second timing question and I consider that in Section VI. A standard fairness hearing for a class-action settlement does not begin to look anything like an antitrust inquiry. There will be no effort to define markets or any effort to inquire systematically into the likely market effects of a settlement. Fairness hearings often will just focus on what the proposed settlement means for the direct parties to the litigation, but even courts that consider more factors, including a public interest factor, are unlikely to conduct a searching antitrust inquiry. The fairness hearing will also not come close to matching the business review process undertaken by the Antitrust Division of the Department of Justice when parties want some level of pre-deal comfort on their planned business arrangements.[4] All of this suggests that the approval of the settlement agreement by the court should not cause some sort of antitrust immunity to attach to the agreement. Under the current caselaw, there is some risk of that and Google and the Authors Guild will clearly argue for such immunities after the fact. The district court considering the agreement might minimize that risk by expressly providing in an order approving the settlement agreement that no antitrust immunities will attach from the court's approval of the agreement—a no *Noerr* clause as it were.[5]

II. Brief Background to the Settlement

Books are the quintessential copyrighted works. The 1790 Copyright Act—the U.S.'s first federal copyright statute—addressed "any map, chart, book or books."[6] And copying a book—in its entirety—is exactly the sort of act that we would think would run afoul of most copyright laws. Of course, a project such as the one envisioned by Google—the world's information online—would necessarily intersect with copyright laws across the planet and across time. To simplify considerably, such a project would necessarily confront three key situations. The first would relate to works in the public domain, that great repository of expression available to be drawn upon by anyone at any time. The second would relate to works of authors or publishers—whomever holds the copyright—who could easily be found. For those works, we might imagine that consent of some sort would be the appropriate vehicle for determining whether works were or were not in our online collection. The need for consent would of course be tempered by the doctrine of fair use which makes possible use of copyrighted works without the copyright holder's permission in circumstances which are, to say the least, unclear.[7] Third, an online database of books would need to figure out what to do about orphan works. These are works that remain within copyright—they have not entered the public domain—but books as to which the copyright holder simply cannot be tracked down. These are not works that we can all draw upon—they are not in the public domain—nor works for which consent provides a simple sorting mechanism.

Notwithstanding all of this, Google pressed forward.[8] After doing preliminary work in 2002 and 2003, on October 7, 2004, Google announced its new Google Print Service at the Frankfurt Book Fair.[9] More than a dozen publishers had agreed to participate in the new service which would bring their books into the Google search engine. Google would provide limited online access to chunks of the books—snippets—while linking to places to buy the books. Two months later, Google announced that it was working with the libraries of Harvard, Stanford, the University of Michigan and the University of Oxford and the New York public library to scan their collections and to bring them online.[10] Michigan made clear that the ambition of the project was to add all of the 7 million volumes in the Michigan library to the Google search engine and to, in the words of University of Michigan president Mary Sue Coleman, launch an era "when the printed record of civilization is accessible to every person in the world with Internet access."[11]

On September 20, 2005, the Authors Guild brought a class action suit against Google alleging copyright infringement relating to the copying of books from the Michigan library.[12] A month later, five publishing companies brought a similar action against Google.[13] Fast forward three years. On October 28, 2008, the Authors Guild, the Association of American Publishers and Google announced a settlement to the pending lawsuits.[14] That settlement agreement, now pending in federal district court in New York, will create a comprehensive new regime for online access to United States books. A fairness hearing will be held on the proposed settlement on October 7, 2009 and, before that date, class members who wish to opt out must do so by September 4, 2009.[15]

III. Five Hypotheticals

It might be useful to frame the GBS settlement by considering five hypothetical cases.

1. *Poodle Book Quote.* An entrepreneur—call the company Poodle—buys a physical copy of every book ever written. Customers call a toll-free number to ask about book quotes. In response to an inquiry, human beings scurry around a vast warehouse of books looking for quotes. Poodle initially charges a modest fee for the service but it is a hit with consumers and, facing no competition, Poodle jacks up its prices, enjoying the benefits of monopoly power.
   What do copyright and antitrust say about this? Nothing, I think. Poodle has purchased books, not made copies of them, and the use that Poodle is making of the books almost certainly falls within traditional notions of fair use. As to antitrust, Poodle has acted on its own and has created a great product with a corresponding market power through successful competition in the marketplace. Antitrust does not condemn this.[16] Indeed, as the Supreme Court put it in its most recent antitrust decision "[s]imply possessing monopoly power and charging monopoly prices does not violate §2 ... ."[17] Tweak this case slightly. Switch from purchased physical copies to digitized copies and have the quotes returned by a computing system algorithmically rather than by human beings. Given the massive copying through digitization, the copyright issues are quite different, but I see no change in the antitrust analysis of the situation.

2. *Digital Book Rights.* An author writes a book, publishes it on paper and retains the copyright. An entrepreneur approaches our author and seeks a license to sell digital copies of the book. Where do we stand? Our author starts with the full set of rights assigned to her by the Copyright Act. Those are statutorily defined rights—start with Section 106—and those rights are limited in some cases by compulsory statutory licenses and by the uncertain but overriding rules of fair use. But none of those rights limit her ability to license whatever rights she has to a third party; indeed, the Copyright Act contemplates such transactions and sets out basic rules governing them.[18] This transaction poses no copyright issues and we should think as such also poses no antitrust issues. Neither copyright nor antitrust insists that an author on her own exploit all of the uses of her work. Put differently, she need not vertically

integrate into all fields where her work might be used. If she prefers to license the right to someone else to exploit her work in a particular medium, she is fully entitled to do so under copyright law and nothing in antitrust should foreclose this. Moreover, copyright law does not create an obligation for her to license her work to a second person merely because she licensed it to a first person. If JK Rowling chooses to allow a movie to be made of her latest Harry Potter novel, she does not need to license all comers who might like a chance to make competing versions of that movie.

3. *Digital Books Cartel.* One hundred authors—all of the authors in our little universe—write novels, publish them on paper and retain copyright to their individual works. They compete vigorously in the offline space with each author setting the price for his work. But as they approach a new medium—digital copies of works—they get together to implement a centralized sale system. In that medium, they set a uniform price for each work of $9.99. What does copyright law or antitrust law say about the situation? Again, copyright law proper says very little about this. Each author would be entitled to exploit her copyright in the new medium. We do see occasional nods towards the doctrine of copyright misuse, which clearly embraces some notion of competitive harm as a within-copyright limitation, but the application of that is typically quite uncertain.[19] But antitrust law is ready to address this situation, as this is classic price-fixing in violation of section 1 of the Sherman Act. Price-fixing remains one of the few behaviors that is *per se* illegal under Section 1.[20] That means that no further inquiry is required into market structure or actual market harm.

4. *Author Book Quote.* Return to the first case, but change the facts. Instead of Poodle buying physical books, the authors/rightsholders get together as a group, digitize copies of their books, and put these online as a searchable quote service. Access to individual books is sufficiently limited that we would not think of the online access as a substitute for purchasing a physical copy of the book. The service is a quote service, with charges per quote or with some sort of unlimited use blanket-license fee.

   Again, there should be no copyright issues here assuming that the authors control their own copyrights. As to antitrust, we are now edging close to something like AS CAP or BMI, where we are nearing seventy years of antitrust regulation of those copyright collectives. Now alter this case slightly and consider a few key questions. They authors conclude that they do not want to enter the book quote business as they do not believe that search is their comparative advantage. But they do form a joint digital rights licensing group with the thought that they will then license those digital rights to firms that want to enter the book quote market. How many licenses would such a monopoly seller want to grant? One to, in my hypo, Poodle? More than one? A license to any entrant in the book quote business? How do we think that book-right licensing would work if the authors could not proceed collectively but instead were required to act individually? Would that alter the number of book quote entrants who would be able to obtain access to some digital rights?

5. *Monopoly by Statute.* Poodle approaches Congress and asks it to enact the "Online Book Quote Monopoly Act of 2009." Under the bill, Poodle would be the only company permitted to offer an online book quote service. Congress passes the act. This would almost certainly be bad policy, but that is not the same thing as an antitrust violation. We have been reluctant to create antitrust roadblocks to efforts to petition the government. Firms can pursue anticompetitive ends through the legal process. We could try to control those efforts through antitrust or we can give free flow to these forces consistent with fundamental First Amendment values. The *Noerr-Pennington* doctrine creates a broad antitrust immunity that protects efforts to seek competitive benefits from the government.[21]

IV. The GBS Settlement

These five cases provide natural starting points for analyzing the settlement agreement currently under consideration in the class-action suit by the Authors Guild against Google for Google's book search service. The actual service provided is substantially more complicated than my examples and the settlement agreement infinitely more complex but we need starting points and these five examples should do the trick.[22]

A. Key Structural Components

We should start by identifying key structural components of the settlement:

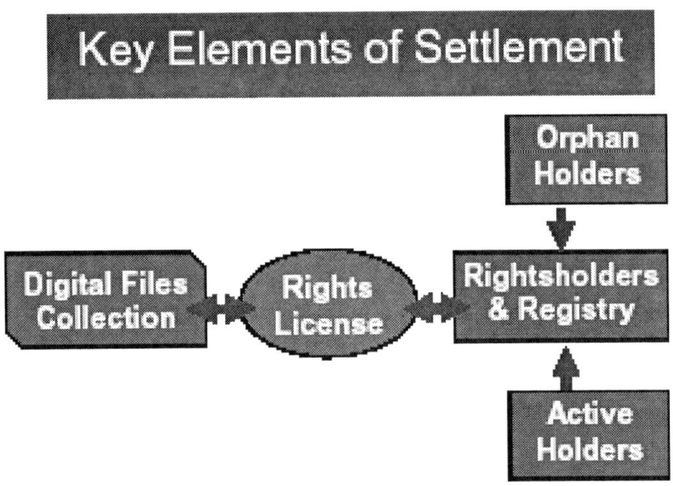

Google has set out to create a collection of digital files. It has done this through scanning books from a variety of sources but the two key sources are those that come from the Google Partner Program and the Google Library Project.[23] This allows Google to create Google Book Search which encapsulates the full variety of uses of the books that it has digitized.[24]

Without licenses from rightholders, it is not clear what Google could do with its digital collection. Google might have litigated a fair use claim to limit its exposure for copyright infringement, but it has instead chosen to seek licenses to use its amassed digital files. The settlement agreement itself operates as a rights license. That license sits between Google, as owner of its digital files collection, and the rightsholders and the registry created by the settlement to act on behalf of the rightsholders. As to the rightsholders, we can usefully divide them into active holders and orphan (inactive) holders.

We might think of the settlement agreement as two related deals neatly fused together. The class action itself is an opt-out class action and therefore any rightsholder who chooses to opt out neither enjoys the benefits of the settlement agreement nor is subject to its terms.[25] Of course, opt out is an active step and the class-action mechanism allows Google to sidestep neatly the problem of orphan works, as the holders of those works presumably will not opt out of the settlement. An orphan holder who shows up and opts out of the settlement is no longer an orphan, as we can now match the right in question with a particular individual or firm.

Active rights holders can embrace the settlement as is or opt out in toto. But opt out is only one way in which the collection can be limited. The agreement contemplates a removal mechanism to remove individual works from Google's collection.[26] The right to remove is time-limited and expires at the end of 27 months after the notice commencement date.[27] There is also a partial removal mechanism, which allows rightholders to exclude a work from particular display uses or revenue models.[28] These mechanisms are substantially more complicated than this quick summary suggests, as the agreement makes an effort to ensure that books made available generally to consumers are also included in institutional subscriptions—the so-called "coupling requirement"—but I will avoid most of these details here.[29]

The definition of "Book" is fundamental both for what it tells us about the works covered by the settlement and for what it says about how Google Book Search will evolve after the settlement is in place.[30] The definition covers written or printed United States works (as defined in the Copyright Act)[31] that have been registered with the U.S. Copyright Office and published before the Notice Commencement Date, which is January 5, 2009, the date of the first notice of the class action settlement.[32] The definition then excludes, among other things, periodicals, music books, works in the public domain and governmental works.

This raises two key points. The settlement is first backwards looking. That is exactly what we would expect for past damages, if any of course, but the settlement also will put in place licenses for the use of these works going forward but only for books that are registered U.S. books before January 5, 2009. That sounds like Google Book Search is not really a library of all books ever written but just those published in the United States before early-2009. A great resource to be sure, but one frozen at a point in time. That takes us naturally to the second point: Google will add content to GBS through separately negotiated contracts. That shows up most directly in the settlement agreement in the Google Partner Program, which contemplates exactly these sort of contracts.[33] To put the point slightly differently, Google must contract going forward to continue to add to its collection and active rightsholders can opt of out of the settlement entirely and instead pursue separate contracts with Google. The group that cannot do that of course are the inactive rightsholders—the holders of rights to the orphan works—and a settlement cut off in early-2009 will encompass all of the preexisting orphan works.

These are the two deals fused together. Active rightholders can effectively embrace simultaneous contracts with Google pursuant to the terms of the settlement or can opt out and seek to execute separate deals with Google. Orphan rightholders will not do anything and the settlement agreement will make it possible for Google, and really only Google, to put those works to use.

That leaves our third building block, the Registry.[34] We have the digital files and the licensing regime that the agreement creates for those files, but the agreement also create a new institution—the Registry—to manage many of the aspects of the settlement agreement. The Registry will act as a middleman between the rightholders and Google. That is both about channeling money but also about managing the information that will be necessary to make this new complicated apparatus work.[35]

B. Use of Digital Copies

We should pay some attention to who gets a digital copy of a book and how it can be used. Google will make a digital copy of a book available to the rightsholder (typically, the author or the publisher).[36] Google will also create a digitized works collection known as the Research Corpus and two or three sites will host it.[37] Libraries that have been the source of the works that make their way into Google's collection will have the chance to receive back a digital copy (the Library Digital Copy).[38] The agreement is a little more complex than that. Cooperating Libraries make books available to Google but do not take back digital copies.[39] In contrast, Fully Participating Libraries receive back digital copies—subject to extensive and complex restrictions—and can make a specified set of uses of those files.[40] The breadth or narrowness of those uses depends on where you sit obviously, but it seems hard to think that these uses, taken individually or together, will amount to meaningful competition to GBS itself.

It is clear that Google thinks of these digital files as such as belonging to Google, as the agreement limits the rights of rightsholders and the Registry to authorize the use of digital copies made by Google.[41] Google is authorized to make Display Uses and Non-Display Uses of the works that make it into GBS.[42] Display Uses turn on the business models embraced in the settlement agreement.[43] Non-Display Uses are at least as interesting, indeed perhaps even more so.[44] Google will be able to draw upon the digitized works to do internal research to improve its core search algorithms—the crown jewel of Google's business—and to develop new services, such as much-improved automatic translation services.

Google's competitors will not fare as well. They might turn to the Research Corpus but the agreement puts substantial limits on the sort of research that can be done. The Research Corpus brings together two overlapping sets of claims, namely, those of Google to the digital files as files it has created through its scanning efforts and then those of the rightsholders to the copyrighted works embedded in the digital files. The Research Corpus is to form the basis for what the agreement calls "Non-Consumptive Research."[45] That is research that is "not research in which a researcher reads or displays substantial portions of a Book to understand the intellectual content presented within the Book." Not reading the book to understand it but instead the use of the book for non- content understanding research. The definition sets forth five examples of research that might qualify, including research on automatic translation;

indexing and search; linguistic analysis and others. This is exactly the sort of research that we should anticipate that Google will perform internally on GBS as part of its right to engage in Non-Display Uses.

The agreement limits the extent to which third parties can do this research. For-profit commercial "use of information extracted from Books" is barred, unless Google and the Registry consent to it.[46] That would seem to prevent the extraction of say, baseball statistics, to provide a fantasy baseball service. Moreover, the agreement expressly limits the use of "data extracted from specific Books" "to provide services to the public or a third party that compete with services offered by the Rightsholders of those books or by Google."[47] That said, commercial exploitation of algorithms developed in doing research on the Research Corpus is permitted.[48] There may be some very fine lines being drawn here. Does algorithmic improvement count as information extracted from books? If so, Google would seem to have the power to block its competitors; if not, the settlement agreement seems to permit this sort of improvement, assuming of course that a prospective researcher can become a "Qualified User."[49] The use of the Research Corpus is limited, in the main, to such users.[50] Google competitors will not easily qualify—researchers based at U.S colleges and universities, non-profits and the government are covered directly—and both the Registry and Google must consent for a for-profit entity to so qualify.

It is not unusual for a firm to condition access to its property in a way that limits subsequent competition. For example, federal patent law makes it possible for a patent holder to limit the assignability of a license that it grants to another person. Absent the limit on assignment, the recipient of a license could immediately compete with the patent holder in the power to deliver a license to a third party. The patent holder would just create a new license for the third party but the original licensee could deliver its license to the third party if licenses were freely assignable. Federal patent law makes it possible for the original patent holder to bar assignment and avoid that competition.[51] To take a second example, courts sometimes enforce limits on reverse engineering of software. The limit on reverse engineering is again intended to limit the ability of the recipient of a work to compete with the originator of that work.[52]

All of that suggests that the limitations imposed by Google on the use of the digital files it has created are broadly consistent with the types of downstream limits on subsequent uses that we see in other circumstances. As to the rightsholders, the only wrinkle is that they are proceeding collectively in limiting that downstream competition. Were the rightsholders to proceed individually rather than collectively, we might very well see more competition as to the uses as to which the works could be put.

Again, this matters most for the orphan works. The settlement agreement is non-exclusive so a downstream user wishing to use a copyrighted work could contract directly with an active rightsholder.[53] The fact that the settlement agreement pushes towards a default position in which the rightsholders will have moved simultaneously in limiting downstream competition may make it easier to limit that subsequent contracting. And of course subsequent contracting is not possible as to the orphan works. But even if we do see direct contracts with active rightsholders, those holders presumably cannot contract as to the digital files that Google has created. The rightsholder can contract as to the copyrighted work, but the digital file itself is a separate matter.[54]

C. Business Models

The settlement agreement contemplates a number of different business models and also contemplates that those business models may change over time. To simplify considerably, focus on institutional subscriptions and consumer purchases. Institutional subscriptions are akin to the blanket licenses that we have seen in ASCAP and BMI. A standard institutional subscription will give access to the entire body of digitized works, but for any particular work in that group, access will be limited. The agreement contemplates a high-transaction cost approach to limiting uses, meaning that it will circumscribe the ability to copy, paste and print. You can get small chunks of the works easily but they will try to make it difficult to aggregate those chunks into something that would compete directly with the traditional offline physical book.[55]

Institutional subscriptions will be priced usually on FTEs—full-time equivalency—meaning, in the case of academic institutions, the number of full- time equivalent students.[56] At what price exactly? The settlement offers pricing objectives that will result simultaneously in the realization of market rates for the books in the collection and in broad access to those books.[57] That may require squaring the circle, but it is clear that many fear that Google will ultimately charge monopoly prices for these institutional subscriptions.[58]

But antitrust law proper may not forbid this. If we treat Google as having as having obtained its monopoly through a risky, entrepreneurial undertaking and therefore legitimately then, as the Supreme Court said most recently in *Linkline*, Google can charge monopoly prices without facing Section 2 liability.[59] Of course, Google's "monopoly" status is seemingly being created by the ability of the rightsholders to act collectively. Were they acting separately, there is every reason to think that we would end up with Google and other firms competing to license from individual rightsholders. That would result in more competition and more fragmentation.

And of course, on this framing, Google would also be the *victim* of the exercise of monopoly power and not just a perpetrator. The rightsholders would have combined horizontally to become the sole source of rights and would have chosen to issue only one license. To be sure, Google would have the right to use the copyrighted works and would be the only firm dealing with end users, but Google presumably would have paid dearly for that monopoly franchise in the deal cut with the rightsholders. If we hold an auction for a monopoly franchise, we will clearly suffer from the downstream harms of monopoly but all of the monopoly profits should be competed away in the auction and should inure to the benefit of the auctioneer, here the rightsholders. The actual situation is even more complex because we think that the auction winner will use what it learns in the direct market in adjacent markets. That is, as noted before, many believe that the Non-Display Uses of the book collection will benefit Google in its core search businesses. The rightsholders would love to access a cut of that incremental value and, again, in an auction process for a single rights franchise, we should expect some of the incremental value in adjacent markets to flow to the rightsholders.

The agreement's second core business model is Consumer Purchases.[60] This seems to contemplate online reading of a particular text with controlled copying, pasting and printing. The pricing mechanism for this access is interesting and seemingly problematic. Books made available this way will be priced either at the price set by the rightsholder or through a mechanism run by Google. The agreement defines a "Settlement Controlled Price."[61] Books will be slotted into particular pricing bins and indeed the agreement contemplates an initial distribution of prices across bins: 5% of the books will be priced at $1.99, 10% at $2.99 and so forth until we reach a final 5% to $29.99.[62] How does this match up with what we think would emerge under standard competition? I am not sure, as that seems to turn on a sense of exactly how spatial competition works, but I do not think that we can be particularly confident that this centralized spacing approach matches what would emerge from normal, decentralized competition.

Then turn to the pricing mechanism itself. Google is to create a “Pricing Algorithm” that Google is to “design to find the optimal such price for each Book and, accordingly, to maximize revenue for each Rightsholder.”[63] That is not what we expect competitive pricing to do. Competition drives down prices to costs and does not have the effect of maximizing revenues to individual competitors. The rightholders are collectively appointing Google as their agent to implement pricing rules for Consumer Purchases that do not seem to mimic what we would see in pure decentralized competition. In that sense, it works its way towards tracking my digital book cartel hypo in Section II above.[64]

The agreement also recognizes that new revenue models might emerge and contemplates that Google and the Registry will negotiate the terms for any new models that might emerge.[65] Finally, the agreement creates a limited, free public access service. That service has been understandably criticized for being quite narrow[66] with one terminal for every 10,000 students at a college and one terminal in public libraries.[67]

D. Who Gets the Money and Unclaimed Funds

Focus next on the money. To track the agreement, start with the big picture and then head into the still-very-important details. Google will pay at least $45 million in cash into the settlement fund, plus another $34.5 million to get the Registry up and running, plus attorneys' fees for the plaintiffs on top of that. On a going forward basis, this is a percentage of revenues deal. Google is to pay the Registry on behalf of the rightsholders 70% of the revenues earned by Google less ten percent for Google's operating costs, resulting in a split of 63% to rightsholders and 37% of revenues to Google.[68]

That is a good starting point but there are some subtleties and they may matter for the overall incentives of the parties. We need to track what the agreement does with regard to revenues associated with public domain works and orphan works. Start with the public domain. The definition of “Book” is one of the key linchpins of the settlement agreement and that definition excludes public domain works.[69] That should have the effect of excluding public domain works from the revenues that would flow to copyright holders. GBS will undoubtedly generate revenues from public domain works and Google will keep all of those revenues. Indeed, the agreement specifically addresses the possibility of mistaken payments by Google with regard to public domain works.[70] All of that certainly matches one vision of how the public domain works: anyone can use it as they wish and make or not make money on it. For fun sometime, go to Amazon and see how many different prices you will find for a work in the public domain.

Turn next to the orphan works. The unclaimed funds provision is particularly important here because of the role that it plays in assigning revenues associated with those works.[71] That provision creates different rules for subscription revenue and non-subscription revenue. Recall that the revenue split in the deal is 70%/30% subject to a 10% discount for Google's operating costs resulting in a net 63%/37% deal. The unclaimed funds provisions make it possible for active rightsholders to get a version of 70% instead of the 63%. Unclaimed funds from non-subscription revenue models are paid first to some of the costs of the Registry and after that to active rightsholders—those who have become Registered Rightsholders—until payments reach 70% of the revenues for their books.[72] That is a 7% carrot and is funded by revenues that arise from orphan works. For revenues that arise from the subscription model, there is a separate Plan of Allocation but, again, unclaimed funds from orphan works will fund additional payments to active rightsholders.[73]

The net effect of the agreement, bolstered by the unclaimed funds provisions, is to turn orphan works into a kind of private public domain. Google will be able to use the orphan works in GBS and both Google and the active rightsholders will benefit from the revenues that arise from those works. This is a common strategy of parties settling intellectual property suits: it is in their joint interest to preserve a property right as against the world. We have seen this pattern before in suits between Lexis and Westlaw regarding the status of page numbers in the West reporting system and in settlements between patent holders and generic entrants in the context of the Hatch-Waxman statute.[74]

V. Multiplying Licenses?

We should turn to the question of how many licenses are likely to be granted to use the orphan works. Recall hypo 4 from Section III where the book authors collectivized and then licensed digital rights to Poodle. I asked then: how many licenses are likely to be granted and how does the fact that the authors are collectivizing influence that? How many licenses should we want to be granted?

A. The Settlement Agreement's Most Favored Nations Clause

To get at this, start with the settlement agreement's most favored nations clause.[75] MFNs are fairly common when deals are done sequentially. An initial contracting party believes that its original agreement will make it possible for its counterparty to enter into other deals with third parties. The original deal may prove the concept and build a blueprint—or at least a starting point—for subsequent deals. An MFN ensures that the beneficiary of it shares in the incremental value that will accrue to its counterparty in other deals and may serve as an important inducement to do the deal in the first place. If potential deal partners believe that there is a second-mover advantage, each partner will hang back and hope that another partner moves first. That logic results in no deals at all—you cannot ever be a second mover if there is a never a first mover. The MFN helps to solve that dynamic problem.

But MFNs can also reduce future competition. A second mover knows that any benefits that it negotiates for will redound to the benefit of the first actor. The second mover knows that it is harder for it to gain a competitive advantage over the first actor and that reduces its incentive to compete to do so in the first place. And for the party granting the MFN, it knows that any benefit it gives to a later contracting party costs it doubly, not just as a cost in the deal with the second actor but also in the original deal with the beneficiary of the MFN. All of this makes clear why MFNs are tricky and probably hard for us to just embrace wholly or seek to forbid.

The actual MFN set out in the Settlement Agreement is interesting.[76] The MFN protects Google from better deals down the road in the standard fashion that MFNs operate. But what is more interesting is that the MFN seems to suggest that the Registry might be able to grant licenses to other third parties. The trigger for the MFN is a deal better than the one that Google is getting in the settlement agreement sometime in the next ten years and that includes "rights granted from a significant portion of Rightsholders other than Registered Rightsholders." That is, a deal about orphan works. But who is going to cut that deal for the orphan works? Google's MFN right is keyed to a deal by the Registry itself or by any substantially similar entity organized by the rightsholders. The MFN certainly operates most naturally in a context where the Registry is understood to have the authority to issue additional licenses for the orphan works. Two other provisions in the settlement agreement might bear on this. The organizational structure provision for the Registry provides that the Registry "will be organized on a basis that allows the Registry, among other things, to ... (iii) to the extent permitted by law, license Rightsholders' U.S. copyrights to third parties."[77] And the settlement agreement extends some authority on its own terms for the Registry to act for rightsholders in executing agreements.[78]

Those get close to the idea that the MFN contemplates—the Registry granting third parties licenses to use the orphan works—but it is not clear that they get there fully. U.S. copyright law does not generally authorize a third party to act on an author's behalf, so it is not clear what the agreement picks up. And the registry authorization clause—a power of attorney-type clause—does not empower the Registry to act directly for orphan authors, unless perhaps we conclude that that idea is embedded itself in the MFN in Section 3.8(a).

B. Clarifying (?) the Settlement Agreement

It seems possible that the settlement agreement intends for the Registry to be able to issue new licenses for the orphan works going forward. Again, it is hard to understand where the current version of Section 3.8(a) will work if the Registry cannot grant new orphan-works licenses. If that is indeed the intent of the settlement agreement, we should clarify 3.8(a) and probably make corresponding changes elsewhere in the agreement, perhaps to 6.2(b)(iii), 6.7 and elsewhere. But I also think that there is a pretty good chance that the current version of the agreement does not contemplate a going-forward licensing power for the orphan works.[79] What should the court do about that in considering whether to approve the settlement agreement?

We should start by assessing the incentives of the rightsholders to voluntarily license their digital works and in so doing also license the rights held by orphan holders. If we think that the collection of rights represented in the lawsuit really is unique, then we should not think that the Authors Guild would wish to license them to a second online book search provider. The rights represent a monopoly and licensing use to two or more providers will result in competition between those providers and will almost certainly make the returns to the rights provider much lower. Indeed, in the extreme case—frequently captured in the notion of Bertrand price competition—we might expect multiple online book search service providers to compete price down to marginal cost which may be close to zero.

If that is right, then the lawsuit by the Authors Guild and the proposed settlement agreement are themselves the vehicle to market power. The opt-out class action increases the likelihood that the rightsholders will act collectively in large numbers by shifting the default position on contracting. This is to take the learning of behavioral economics on the importance of defaults and turn it into large-scale contracting to achieve market power. That probably works that way even for identifiable rightsholders but it clearly operates in that way as to orphan authors. Nothing would have prevented large numbers of rightsholders from entering into private contracts with Google to create something akin to GBS but the numbers—and the resulting market power—would clearly have been smaller without the switch in default settings made by the opt-out class action. And what those private contracts could not have accomplished was bringing the orphan works into the deal, but the opt-out class action settlement does just that.

We would seem to have two natural approaches to changing the settlement agreement to ensure the possibility of competing digital books collections: (1) alter the settlement agreement now as part of the approval process to add additional licenses; or (2) ensure that the agreement contains a mechanism for new licenses to be considered going forward. Start with what those licenses might look like substantively. The current MFN gives Google the benefit of any other deals in the next ten years for new licenses of the orphan works. A fully symmetric MFN would make it possible for other entities to get the licensing regime that Google gets for the works licensed by active rightsholders and for the orphan works.

As to the latter, Google can use those fully and the release provisions in the settlement agreement ensure that Google will not face liability for copyright infringement for its use of those works.[80] Google will make payments to the Registry for the revenues that it derives from those works, just as it would any other work in GBS. And the Registry in turn distributes those revenues per the terms of the Settlement Agreement. Remember importantly that active rightsholders effectively benefit from the revenues generated by the orphan works.

A symmetric MFN would allow qualified entities to sign up for the going-forward provisions of the settlement agreement as to both the works of the active rightsholders and the orphan works. That license would mitigate both the problem that the active rightsholders will be reluctant to issue additional licenses on their own for their works—why compete with themselves?—and the problem that the settlement represents the only way to gain access to the orphan works. Note of course that those rights licenses would not actually enable immediate entry into the book search business. Any entrant will have to do its own scanning as nothing I have said here would justify some sort of mandatory access to the scans that Google has created. We should want to foster competition in scanning.

What does "qualified" mean? The rightsholders undoubtedly will emphasize that they cut a deal with Google, not any random entity. The rightsholders are relying on Google's incentive to go out and make money and pay 63% of the revenues generated to the rightsholders. A non-profit would not have the same incentives to generate revenues. And the rightsholders will argue that they have negotiated for an elaborate protective scheme for the scans that Google has created.[81] the rightsholders will understandably want a Google-class contracting partner if we are going to force them to duplicate the settlement agreement licenses and make those available to others. But those limits should not mean that non-profits are completely left out. The revised MFN should separate out the orphan-works deal from the deal made by the active rightsholders—unbundle the orphan-works license from the overall settlement—and make corresponding licenses (and releases) available to third- parties to use the orphan works.[82]

Active rightsholders might very well object to the Xeroxing of their deal to allow other firms access to licenses to use their works. But that would almost certainly reflect a desire to preserve the incremental market power that accrues to them from the ability to implement their deal with Google through the court system. Proceeding through the court system maximizes their ability to act collectively and it is that which shrinks the number of licenses granted to their works. If the active rightsholders were forced to act outside the court system, we would almost certainly see greater fragmentation in the licenses granted and more competition in book search.

And the active rightsholders would seem to have little basis to object to the expansion of the MFN to encompass additional licenses of the orphan works. I understand that they might want to squelch the competition that might emerge if multiple firms had access to the orphan works. That competition could easily reduce overall industry revenues for digital book search services and that would be against the interest of the active rightsholders given that they are cutting deals tied to percentages of revenues. But eliminating competition cannot be seen as a legitimate goal of the collective action of the active rightsholders captured in the settlement agreement. Absent the lawsuit, the active rightsholders could convey no rights in the orphan works to anyone and we need to guard against their ability to create a monopoly as to those works by issuing only a single license to those works.

The case against expanding the scope of the MFN now or ensuring a mechanism to issue additional licenses going forward might run as follows. Google undertook a substantial risk in digitizing works and putting them online. Even in proceeding carefully by controlling the scope of access to the works, Google faced substantial liability for the core copying of entire works that made this possible. To be sure, Google could make arguments regarding the permissibility of intermediate copying and regarding fair use, but those arguments were far from a sure thing, plus Google was operating at a sufficient scale that were it to lose the damages would likely be substantial.

In granting access to third parties to the rights regime created by the settlement agreement, we face a standard selection problem in that potential competitors are always eager to join in successful projects and share those costs, but rarely volunteer to fund those which have been revealed to be losers. That means that any notion of merely paying some measure of Google's costs in creating the new licensing regime is inadequate in that those costs need to be grossed up for the risk of failure.

If the settlement represents a clever solution to the orphan works problem, we could imagine a number of different approaches to follow-on efforts. One approach would require a subsequent entrant to pursue the same legal strategy. Of course, the path here was entry by Google followed by a class action lawsuit by the Authors Guild. As suggested before, it is not clear that the Authors Guild will want to grant a second collective rights license and that means a second suit might not ever be brought. Of course, the entrant might bring a declaratory judgment action naming as defendants the plaintiff class in the current litigation. But we should ensure that a subsequent entrant does not face an entry barrier based on the inability of a court to obtain new jurisdiction vis-à-vis the orphan works or based on the unwillingness of active rightsholders to grant a second license to the orphan works.

That would suggest a possible more limited approach to the settlement agreement: do not create additional licenses in the settlement agreement now but make sure that some combination of the Registry and the court can do so going forward. As noted before, there is a reading of the settlement agreement that suggests that the Registry can grant new licenses to orphan works under the current agreement. And we could make that more explicit in the agreement. Alternatively—or perhaps in addition—the settlement agreement provides that the district court will retain jurisdiction going forward.[83] If that retained jurisdiction was understood to cover efforts by qualified firms to license works covered by the settlement agreement and to license separately the unbundled orphan works—perhaps thought of as intervening defendants in the original case—that might solve the fear that the Authors Guild might not sue again or that jurisdiction might not attach for a declaratory judgment action.

VI. Antitrust Immunities and the Timing of Antitrust Inquiry

To get a handle on the question of timing, consider a counterfactual. Suppose that Google launched its service as it did and that the Authors Guild drafted a complaint identical to the one filed. But in this alternative universe, no lawsuit is filed. Google and the Authors Guild negotiate an arrangement identical to the settlement agreement but they do so without the intervention of litigation. Where would we stand and how does that differ from the current situation?

A number of points matter. The arrangement between Google and the Authors Guild would simply be a matter of contract. Rightsholders who signed on the dotted line would become bound by the contract. Other rightsholders who declined the agreement would of course not be bound. No federal district court judge would have any role in approving the agreement. This would be purely a private matter and there would be no substantive review of the contract at the time of its execution. There would be no contemporaneous evaluation of whether the deal was fair as between the parties or what antitrust consequences might be of the new arrangement. There would of course be no possibility that some sort of *Noerr-Pennington* immunity would attach to the contract. Nor would there be a consideration of whether the contracting rights holders were somehow sufficiently similar that they could sign the same contract.

Now revert to the actual case. A lawsuit was filed: how does that change the analysis? The key point of course is that the nature of an opt-out class action means that the default setting as to whether or not the settlement is agreed to has changed. If a rightsholder does not affirmatively decline the contract, he or she is bound. That is the precise flip of the position we had in the counterfactual, and default positions, as behavioral economics is quick to point out, can matter enormously.

We also will have the fairness hearing on the settlement, but the antitrust analysis in the settlement hearing on the class action is likely to be minimal to non-existent. Circuits differ in the factors that they consider in a fairness hearing, with some looking to many factors, including the public interest, with others—and probably the Second Circuit—focusing more narrowly on what the settlement means for the parties to the lawsuit.[84]

Now we can frame the *Noerr-Pennington* point. In circuits that embrace a party-centered approach to fairness and class-action settlements, no possibility of subsequent *Noerr-Pennington* immunity for the court-approved agreement should attach. The court will not have considered what the competitive consequences would be of the arrangement and therefore clearly have engaged in no shaping or assessing of the agreement in those terms.[85] This case should be treated as our counterfactual case would be where no complaint had been filed. That private contract enjoyed no immunity from antitrust inquiry and in similar fashion these court-approved agreements should not either. And even in circuits that embrace a broader inquiry to fairness hearings, the antitrust analysis is likely to be minimal. This suggests, again, that no *Noerr-Pennington* immunity should attach to the approval of these settlements.[86] If the district court approves the settlement agreement, it should take the additional step of including a no *Noerr* clause in its order approving the settlement. That clause would provide that no antitrust immunities would attach to the settlement agreement from the court's approval of it.

VII . Conclusion

Google boldly launched Google Book Search in pursuing its goal of organizing the world's information. Even though Google was sensitive to copyright values, the service relied on mass copying and thus Google undertook a substantial legal risk in setting up the service. That risk was realized with the lawsuits by the Authors Guild and the Association of American Publishers. The October, 2008 settlement agreement for those suits will create an important new copyright collective and will legitimate broad-scale online access to United States books registered before early January, 2009.

The settlement agreement is exceedingly complex but I have focused on three issues that raise antitrust and competition policy concerns. First, the agreement calls for Google to act as agent for rightsholders in setting the price of online access to consumers. Google is tasked with developing a pricing algorithm that will maximize revenues for each of those works. Direct competition among rightsholders would push prices towards some measure of costs and would not be designed to maximize revenues. As I think that that level of direct coordination of prices is unlikely to mimic what would result in competition, I have real doubts about whether the consumer access pricing provision would survive a challenge under Section 1 of the Sherman Act.

Second, and much more centrally to the settlement agreement, the opt-out class action will make it possible for Google to include orphan works in its book search service. Orphan works are works as to which the rightsholder cannot be identified or found. That means that a firm like Google cannot contract with an orphan holder directly to include his or her work in the service and that would result in large numbers of missing works. The opt-out mechanism—which shifts the default from copyright's usual out to the class action's in—brings these works into the settlement.

But the settlement agreement also creates market power through this mechanism. Absent the lawsuit and the settlement, active rightsholders could contract directly with Google, but it is hard to get large-scale contracting to take place and there is, again, no way to contract with orphan holders. The opt-out class action then is the vehicle for large-scale collective action by active rightsholders. Active rightsholders have little incentive to compete with themselves by granting multiple licenses of their works or of the orphan works. Plus under the terms of the settlement agreement, active rightsholders benefit directly from the revenues attributable to orphan works used in GBS.

We can mitigate the market power that will otherwise arise through the settlement by expanding the number of rights licenses available under the settlement agreement. Qualified firms should have the power to embrace the going-forward provisions of the settlement agreement. We typically find it hard to control prices directly and instead look to foster competition to control prices. Non-profits are unlikely to match up well with the overall terms of the settlement agreement, which is a share-the-revenues deal. But we should take the additional step of unbundling the orphan works deal from the overall settlement agreement and create a separate license to use those works. All of that will undoubtedly add more complexity to what is already a large piece of work, and it may make sense to push out the new licenses to the future. That would mean ensuring now that the court retains jurisdiction to do that and/or giving the new Registry created in the settlement the power to do this sort of licensing.

Third, there is a risk that approval by the court of the settlement could cause antitrust immunities to attach to the arrangements created by the settlement agreement. As it is highly unlikely that the fairness hearing will undertake a meaningful antitrust analysis of those arrangements, if the district court approves the settlement, the court should include a clause—call this a no *Noerr* clause—in the order approving the settlement providing that no antitrust immunities attach from the court's approval.

End Notes

[1] *See* Google Corporate Information, http://www.google.com/corporate/ (last visited July 8, 2009).

[2] *See* The Official Settlement Agreement, http://www.googlebooksettlement.com/ (last visited July 8, 2009). *See also* Google's Discussion of the Settlement, http://books.google.com/googlebooks/agreement/ (last visited July 5, 2009).

[3] *See* James Gibson, *Google's New Monopoly*, Wash. Post, Nov. 3, 2008, at A21; *see* James Grimmelmann, *How to Improve the Google Book Search Settlement*, 12 J. Internet L. at 1, 11-20 (April 2009), *available at* http://works.bepress.com/cgi/viewcontent.cgi?article=1022&context=james_grimmelmann; *see* Robert Darnton, *Google & the Future of Books*, 56 N.Y. Rev. Books 2 (Feb. 12, 2009), *available at* http://www.nybooks.com/articles/22281.

[4] 28 C.F.R. § 50.6 (2009). *See also* Antitrust Division, U.S. Department of Justice, Instructions for Business Reviews, http://www.usdoj.gov/atr/public/busreview/201659b.htm (last visited July 5, 2009).

[5] *See* Professional Real Estate Investors, Inc. v. Columbia Pictures Industries, Inc., 508 U.S. 49 (1993); Eastern Railroad Presidents Conference v. Noerr Motor Freight, Inc., 365 U.S. 127 (1961); United Mine Workers v. Pennington, 381 U.S. 657 (1985).

[6] 1 Stat. 124 (1790).

[7] *See* 17 U.S.C. § 107 (2009).

[8] *See* History of Google Book Search, http://books.google.com/googlebooks/history.html (last visited July 5, 2009).

[9] *See* Edward Wyatt, *New Google Service May Strain Old Ties in Bookselling*, N.Y. Times, Oct. 8, 2004, *available at* http://www.nytimes.com/2004/10/08/technology/08book.html.

[10] *See* Press Release, Google Inc., Google Checks Out Library Books (Dec. 14, 2004), http://www.google.com/press/pressrel/print_library.html.

[11] *See* Press Release, University of Michigan News Service, Google/U-M Project Opens the Way to Universal Access to Information (Dec. 14, 2004), http://www.umich.edu/news/?Releases/2004/Dec04/library/index.

[12] *See* Class Action Complaint, The Author's Guild v. Google Inc., 05 CV 8136 (S.D.N.Y. 2005), *available at* http://www.authorsguild.org/advocacy/articles/settlement-resources.attachment/authors-guild-v-google/Authors%20Guild%20v%20Google%2009202005.pdf.

[13] *See* Complaint, McGraw-Hill Co., Inc. v. Google Inc., 05 CV 8881 (S.D.N.Y. 2005), *available at* http://www.authorsguild.org/advocacy/articles/settlement-resources.attachment/mcgraw-hill/McGraw-Hill%20v.%20Google%2010192005.pdf.

[14] *See* Press Release, The Authors Guild, Authors, Publishers and Google Reach Landmark Settlement (Oct. 28, 2008), http://www.authorsguild.org/advocacy final_ 102808 /press_release_final_102808.pdf.

[15] *See* Authors Guild v. Google Settlement Resources Page, http://www.authorsguild. org/advocacy/ articles/ settlement-resources.html (last visited July 5, 2009).

[16] As Judge Learned Hand famously put it in United States v. Aluminum Co. of America, 148 F.2d 416, 430 (1945): "A single producer may be the survivor out of a group of active competitors, merely by virtue of his superior skill, foresight and industry. In such cases a strong argument can be made that, although the result may expose the public to the evils of monopoly, the Act does not mean to condemn the resultant of those very forces which it is its prime object to foster: finis opus coronat. The successful competitor, having been urged to compete, must not be turned upon when he wins."

[17] Pacific Bell Telephone Co. v. Linkline Communications, Inc., 555 U.S. ___ (2009).

[18] *See* 17 U.S.C. § 201-04 (2009).

[19] *See, e.g.,* Practice Management Information Corp. v. American Medical Ass'n, 121 F.3d 516 (9th Cir. 1997), *amended by* 133 F.3d 1140 (9th Cir. 1998).

[20] NYNEX Corp. v. Discon, Inc., 525 U.S. 128, 133 (1998) (citing cases).

[21] *See supra* note 5.

[22] And the settlement agreement itself is not the only relevant document to the rights that will emerge from this process. For example, subsequent to completing the settlement agreement, Google signed a revised digitization agreement with the University of Michigan. *See* Press Release, University of Michigan News Service, U-M first to sign new digitization agreement with Google (May 20, 2009), http://www.ns.umich.edu/htdocs/releases/story.php?id=7162.

[23] *See* Settlement Agreement at §§ 1.62, -61, The Authors Guild, Inc., Association of American Publishers, Inc., et al. v. Google Inc., No. 05 CV 8136 (S.D.N.Y. Oct. 28, 2008), *available at* http://www.authorsguild.org/advocacy 20Agreement.pdf ("Settlement Agreement").

[24] *Id.* §§ 1.60, 2.1(a).

[25] *Id.* § 17.33.

[26] *Id.* §§ 1.124, 3.5(a).

[27] *Id.* § 3.5(a)(iii).

[28] *See* Settlement Agreement, *supra* note 23, at 3.5(b).

[29] *Id.* § 3.5(b)(iii).

[30] *Id.* § 1.16.

[31] *See* 17 U.S.C. § 101 (2009).

[32] *See* Settlement Agreement, *supra* note 23, at § 1.94.

[33] *Id.* § 1.62.

[34] *Id.* § 1.123, Art. VI.

[35] *Id.* §§ 6.1(d), 6.3, 6.5, 6.6.

[36] *Id.* § 3.11.

[37] *See* Settlement Agreement, *supra* note 23, at §§ 1.130, 7.2(d).

[38] *Id.* § 1.78.

[39] *Id.* § 1.36.

[40] *Id.* §§ 1.58, 7.2(a), 7.2(b), 7.2(c).

[41] *Id.* §§ 3.12, 6.6(b).

[42] *See* Settlement Agreement, *supra* note 23, at § 2.2.

[43] *Id.* § 1.48.

[44] *Id.* § 1.91; *see* Fred von Lohmann, *Google Book Search Settlement: A Reader's Guide* (2008), http://www.eff.org/deeplinks/2008/10/google-books-settlement-readers-guide.

[45] *See* Settlement Agreement, *supra* note 23, at § 1.90.

[46] *Id.* § 7.2(d)(viii).

[47] *Id.* § 7.2(d)(ix).

[48] *Id.* § 7.2(d)(x).

[49] *Id.* § 1.121.

[50] *See* Settlement Agreement, *supra* note 23, at § 7.2(d)(iii).

[51] Everex Systems, Inc. v. Cadtrak Corp., 89 F.3d 673, 679 (9th Cir. 1996).

[52] Bowers v. Baystate Technologies, Inc., 320 F.3d 1317, 1324-26 (Fed. Cir. 2003).

[53] *See* Settlement Agreement, *supra* note 23, at § 2.4.

[54] It is not clear to me whether the settlement agreement makes a Host Site an owner of the Research Corpus. There is a mechanism for removing works from the Research Corpus, *id.* § 7.2(d)(iv), but that could just mean that the Host Site holds title, but a defeasable one, to the copy in the question. What turns on this could be the application of the first-sale doctrine, 17 U.S.C. § 109(a) (2009), though that doctrine seems to contemplate sale

or other dispossession of the copy in question and would not seem to free the Host Site from the licensing limits of the settlement agreement.

[55] *See* Settlement Agreement, *supra* note 23, at § 4.1(d).

[56] *Id.* § 4.1(a)(iii).

[57] *Id.* § 4.1(a)(i).

[58] As Harvard Librarian Robert Darnton has put it "... Google will enjoy what can only be called a monopoly—a monopoly of a new kind, not of railroads or steel but of access to information." Google will "first, entice subscribers with low initial rates, and then, once they are hooked, ratchet up the rates as high as the traffic will bear." Darnton *supra* note 3.

[59] *See supra* note 17.

[60] *See* Settlement Agreement, *supra* note 23, at §§ 1.32, 4.2.

[61] *Id.* § 1.143, 4.2(b)(i)(2).

[62] *Id.* § 4.2(c)(ii).

[63] *Id.* § 4.2(b)(ii).

[64] Two new papers develop this point further. *See* Eric M. Fraser, Antitrust and the Google Books Settlement: The Problem of Simultaneity (June 23, 2009) (unpublished manuscript, on file with author); *see* Christopher Suarez, Proactive FTC/DOJ Intervention in the Google Book Search Settlement: Defending Our Public Values, Protecting Competition (May 25, 2009) (unpublished manuscript, available at http://ssrn.com/abstract=1409824).

[65] *See* Settlement Agreement, *supra* note 23, at § 4.7.

[66] Jennifer Howard, *Harvard Says No to Google Deal for Scanning In-Copyright Works*, Chron. Higher Educ., Oct. 20, 2008, *available at* http://chronicle.com/news/article/5417/harvard-says-no-thanks-to-google-deal-for-scanning-in-copyright-works.

[67] *See* Settlement Agreement, *supra* note 23, at §§ 4.8(a)(i)(1), 4.8(a)(i)(3).

[68] *Id.* § 2.1(a). To track the agreement with more detail, Section 4.5(a) defines a standard revenue split for purchases as 70% of Net Purchase Revenues. The definition of that in turn, *id.* 1.87, makes clear that those are all revenues received by Google from all revenue models other than advertising, less the 10% operating cost deduction. There are parallel provisions for advertising revenues, id. §§ 4.5(a)(ii), 1.86, though the fact that there are parallel provisions rather than one unified provision suggests that I may be missing some subtle difference between the two.

[69] *See* Settlement Agreement, *supra* note 23, at § 1.16.

[70] *Id.* § 6.3(b).

[71] *Id.* § 6.3(a).

[72] *Id.* § 6.3(a)(i).

[73] *Id.* § 6.3(a)(ii), Attach. C, 1.1(e).

[74] *See* Stephen Labaton, *Westlaw and Lexis Near Truce*, N.Y. Times, July 19, 2008 *available at* http://www.nytimes.com/1988/07/19/business/westlaw-and-lexis-near-truce.html; *see* In re Cardizem CD Antitrust Litigation, 332 F.3d 896 (6th Cir. 2003).

[75] *See* Settlement Agreement, *supra* note 23, at § 3.8(a).

[76] *Id.* § 3.8(a) ("Effect of Other Agreements. The Registry (and any substantially similar entity organized by Rightsholders that is using any data or resources that Google provides, or that is of the type that Google provides, to the Registry relating to this Settlement) will extend economic and other terms to Google that, when taken as a whole, do not disfavor or disadvantage Google as compared to any other substantially similar authorizations granted to third parties by the Registry (or any substantially similar entity organized by Rightsholders that is using any data or resources that Google provides, or that is of the type that Google provides, to the Registry relating to this Settlement) when such authorizations (i) are made within ten (10) years of the Effective Date and (ii) include rights granted from a significant portion of Rightsholders other than Registered Rightsholders. ...").

[77] *Id.* § 6.2(b).

[78] *Id.* § 6.7 ("Authorization of Registry. Where this Settlement Agreement confers on the Registry rights and obligations with respect to Books and Inserts, including with respect to the Registry's relationship with each of Google, the Fully Participating Libraries, the Cooperating Libraries and the Public Domain Libraries, Plaintiffs and all Rightsholders, as of the Effective Date, shall be deemed to have authorized the Registry to exercise such rights and perform such obligations on behalf of the Rightsholders with respect to their respective Books and Inserts, including to enter into Library-Registry Agreements.").

[79] Google has addressed this question on its public policy blog:

> Under the settlement Google will be able to open up access to truly orphaned books, but we still think more needs to be done to allow anyone and everyone to use these works. Any company or organization that wants to open up access to this untapped resource should be able to do so. The settlement is not a panacea, since it only covers a subset of orphaned works, provides only certain uses, and is not able to extend these uses to other providers. The need for comprehensive orphan works legislation is not diminished.

Derek Slater, Google Book Search settlement and access to out of print books, June 2, 2009, http://googlepublicpolicy.blogspot.com/2009/06/google-book-search-settlement-and.html.

[80] *Id.* at Art. X.

[81] *Id.* at Art. VIII, Attach. D.

[82] One could imagine any number of possible non-profit entrants, including the Internet Archive, the HathiTrust, JSTOR or a completely new nonprofit devoted to these issues.

[83] *Id.* § 17.23.

[84] As the Sixth Circuit framed its test in UAW v. General Motors Corp., 497 F.3d 615, 631 (6th Cir. 2007): "Several factors guide the inquiry: (1) the risk of fraud or collusion; (2) the complexity, expense and likely duration of litigation; (3) the amount of discovery engaged in by the parties; (4) the likelihood of success on the merits; (5) the opinions of class counsel and class representatives; (6) the reaction of absent class members; and (7) the public interest."

[85] This is likely to be true, even if the Antitrust Division of the Deparment of Justice makes a filing in the case, as it may well do. *See* Letter from William G. Cavanaugh to Judge Denny Chin (July 2, 2009) (available at Order at 2, The Authors Guild, Inc., Association of American Publishers, Inc., et al. v. Google Inc., No. 05 CV 8136 (S.D.N.Y. July 2, 2009)).

[86] Like most interesting propositions, the caselaw is not clear on whether *Noerr-Pennington* immunity would attach to the court's approval of the settlement agreement. The Supreme Court certainly did not allow a court consent decree to block additional antitrust inquiry into the arrangements in ASCAP and BMI. See Broadcast Music, Inc. v. Columbia Broadcasting System, Inc., 441 U.S. 1, 13 (1979) ("Of course, a consent judgment, even one entered at the behest of the Antitrust Division, does not immunize the defendant from liability for actions, including those contemplated by the decree, that violate the rights of nonparties."). But more recent lower court decisions have clearly relied on *Noerr-Pennington* to block some challenges to court-approved arrangements. See Sanders v. Brown, 504 F.3d 903, 913 n.8 (9th Cir. 2007) (finding that *Noerr-Pennington* doctrine protected a private party from antitrust liability from anti-competitive harms resulting from tobacco settlement agreement negotiated with state entity but further declining to resolve whether that doctrine would insulate private parties from an anti-competitive settlement merely because that agreement was approved by a court). The Second Circuit has taken a narrower approach than most circuits to the scope of *Noerr-Pennington* immunity in the master settlement agreements for the tobacco litigation. *See* Freedom Holdings, Inc. v. Spitzer, 357 F.3d 205 (2nd Cir. 2004).

In: Digital Books: Competition and Commerce
Editor: Oliver A. Hagen ISBN: 978-1-61728-299-7

Chapter 9

STATEMENT OF THE NATIONAL FEDERATION OF THE BLIND, BEFORE THE COMMITTEE ON THE JUDICIARY, HEARING ON "COMPETITION AND COMMERCE IN DIGITAL BOOKS"

Marc Maurer

Mr. Chairman: My name is Marc Maurer, and I serve as President of the National Federation of the Blind. The National Federation of the Blind is the oldest and largest organization of blind people in the United States. Through our affiliates in each of the fifty states, the District of Columbia, and Puerto Rico, and our 700 local chapters, we seek to advance the rights of blind people by helping both the blind and the sighted to understand that blindness, in and of itself, need not be a tragedy. The real problem of blindness is not loss of eyesight, but misunderstandings and misconceptions about it that are prevalent in society. With proper training and opportunity, blindness can be reduced to the level of a mere physical nuisance.

The "proper training and opportunity" part of this equation cannot be overemphasized. Among the keys to opportunity are a good education and access to information. This is particularly true in the twenty-first century, when information is the coin of the realm. In order to succeed, blind people must have full and equal access to all forms of information, including the written word. But historically, access to written knowledge has been one of the greatest challenges faced by the blind. For most of human history, the vast majority of blind people did not have access to books at all. With the invention of Braille, and then of audio recording and playback technology, more books became available to us, but still the selection of titles to which we have access is extremely limited.

The Google settlement, if approved, will change all that. The agreement promises to make millions of titles available to a wide audience, including those who are blind or who cannot read print for other reasons. The terms of the settlement allow Google to provide the material it offers users "in a manner that accommodates users with print disabilities so that such users have a substantially similar user experience as users without print disabilities." A user with a print disability under the agreement is one who is "unable to read or use standard printed material due to blindness, visual disability, physical limitations, organic dysfunction, or dyslexia." The settlement, if approved, will therefore bring the printed word to as many as thirty million people who currently have limited access to it. Blind people will be able to search for books through the Google Books interface and purchase, borrow, or read at a public library any of the books that are available to the general public, in a format that is compatible with text enlargement software, text-to-speech screen access software, and refreshable Braille devices. If this settlement is approved, blind people will have greater access to books than we have ever had in human history.

Currently, the blind have access to a relatively sparse selection of titles produced by government and nonprofit organizations dedicated to serving our needs. The primary source of reading material for most blind Americans is the National Library Service for the Blind and Physically Handicapped of the Library of Congress. While this service has done an outstanding job of providing books to the blind within budgetary constraints, it has, at most, perhaps 70,000 circulating titles in its collection. Furthermore, even if all methods of providing content in a format accessible to the blind, including commercial audio books, are taken into consideration, it is estimated that only 5 percent of the books published each year are ever produced in a format that can be used by the blind or others with print disabilities.

While the blind have for years been tantalized by the promise of greater access through e-books, which are inherently accessible, the sad fact is that no other provider of e-books has yet made its offerings accessible to the blind. Instead, publishers distribute their e-books through inaccessible devices and platforms, use digital rights management schemes to prevent screen access technology and other methods used by the blind from accessing these books, or both. Prior to this landmark settlement agreement, neither authors and publishers, nor any entity promoting e-book technology had ever consented to any system that would make such a large number of books immediately accessible to the blind and other Americans with print disabilities. The Google settlement therefore represents the only present prospect for blind people to access electronic texts. It is our hope and belief that the settlement will also represent a paradigm shift that will encourage other publishers and e-book distributors to make accessibility a priority.

Libraries are currently distributing e-books to borrowers. An increasing number of universities require students to use e-texts. Some state governments are contemplating replacing printed books with e-texts. A number of companies are distributing e-texts. The e-text market is expanding rapidly, and the Google settlement represents the only substantial effort to make this form of information usable by the blind. The technology exists to make all of this distributed information readily accessible to the blind. Unless an immediate effort to implement such a system is pursued, the result for employment for blind people will be devastating. Already the lack of information has meant that fewer than 50 percent of blind people in their school years will graduate from high school. Already the lack of information has contributed to an unemployment rate for the blind in excess of 70 percent. Already the lack of information has prevented blind students from being able to matriculate in college courses of their choosing. Blind people are willing to pay for books, but they must be books we can read. The proposed Google settlement is a promise that this magnificent notion may become real.

It has been said that the proposed Google settlement is an effort to give judicial sanction to theft of intellectual property. This argument ignores the provisions of the settlement itself.

Any author wishing to avoid the terms of the settlement may opt out. Those who do not opt out receive a very substantial portion of the revenue generated.

Some opponents of the settlement have propounded the argument that by including accessible provisions for the blind in the agreement Google has taken a position akin to that of "Robin Hood." Apparently, these opponents are saying that Google is taking what doesn't belong to it and justifying the theft by giving access to the blind. This is an argument without foundation. The blind are prepared to pay for books at the same rate and to the same extent that the sighted must do so. The settlement has no provisions to give intellectual property to the blind that is not already available to the sighted. The blind must pay as much for intellectual property as the sighted must. No different terms apply to blind people from those which apply to the sighted. The "Robin Hood" analogy fails because the blind don't get anything for free that the sighted don't get for free and because a "Robin Hood" must first steal the material. If the theft hasn't occurred, the transaction has to be characterized in some other way.

The Google settlement is, for the blind and many others, the next step in the democratization of knowledge. That process began with the introduction of the printing press and then, for the blind, with the invention of Braille. Now technology has arisen which transcends the traditional limitations of both print and Braille, promising to make millions of titles available to the blind in Braille or any other format of our choice. The narrow business interests of Google's competitors must not be allowed to block Americans who cannot read print from all of the opportunities that greater access to written knowledge will make available to them. It is time for the doors of the world's great libraries to be opened in welcome to everyone.

Thank you for the opportunity to present this testimony. If you have any further questions, please do not hesitate to contact the National Federation of the Blind.

In: Digital Books: Competition and Commerce
Editor: Oliver A. Hagen
ISBN: 978-1-61728-299-7

Chapter 10

TESTIMONY OF JOHN M. SIMPSON, CONSUMER ADVOCATE WITH CONSUMER WATCHDOG, HEARING ON "COMPETITION AND COMMERCE IN DIGITAL BOOKS"

John M. Simpson

Good morning Chairman Conyers, Ranking Member Smith, and members of the committee. My name is John M. Simpson and I am a consumer advocate with the nonprofit, nonpartisan Consumer Watchdog, formerly the Foundation for Taxpayer and Consumer Rights. Thank you for considering my testimony.

Established in 1985, Consumer Watchdog is a nationally recognized non-partisan, non-profit organization representing the interests of taxpayers and consumers. Our mission is to provide an effective voice for the public interest. Consumer Watchdog's programs include health care reform, oversight of insurance rates, energy policy, protecting civil justice, corporate reform, and political accountability. Over the past year our Google Privacy Project, funded by The Rose Foundation, a charitable nonprofit organization, has focused on Google's privacy policies and sought better privacy guarantees from the company for users of its services.

During this project we became aware of the proposed Google Books settlement and were concerned about its impact on consumer privacy and anticompetitive aspects. In April we called upon the Department of Justice to intervene in the proposed settlement because of antitrust concerns and Justice has since announced it is investigating.

Let me absolutely clear. We do not oppose the concept of digital libraries. Done correctly, they would greatly enhance public access to books. Everyone should be in favor of that. I also personally believe, like Google, that in the digital age scanning books to convert them to a digital database and then displaying "snippets" in a search of that database is appropriate fair use. However, that understanding has not been substantiated by a court and was the initial activity that prompted the suit against Google resulting in the proposed settlement.

The problem is Google's monopolistic digital library and how it would be implemented. As we said in our friend-of-the-court brief in the Google Books case, "The proposed class-action settlement is monumentally overbroad and invites the Court to overstep its legal jurisdiction, to the detriment of consumers and the publicÉThe proposed Settlement Agreement would strip rights from millions of absent class members, worldwide, in violation of national and international copyright law, for the sole benefit of Google."

The parties in the suit negotiated the Google Books settlement in secret and there was no opportunity to represent and protect the broad interests of all consumers. This deal simply furthers the relatively narrow agenda of Google, The Authors Guild and the Association of American Publishers.

The settlement provides a mechanism for Google to deal with "orphan works." Orphan works are works under copyright, but with the rights holders unknown or not found. The danger of using such works is that a rights holder will emerge after the book has been exploited and demand substantial infringement penalties. The proposed settlement protects Google from such potentially damaging exposure, but provides no protection for others. This effectively is an insurmountable barrier for potential competitors who wish to enter the digital book business.

Consumer Watchdog asked U.S. District Court Judge Denny Chin to reject the proposed settlement for four reasons:

- **It is not fair, adequate or reasonable because it far exceeds the actual controversy before the court and abuses the class action process:** "The proposed class action settlement claims to resolve the actual dispute between the parties, but it also goes much, much farther, and purports to enroll millions of absent class members in a series of new business 'opportunities.' For those absent class members who fail to step forward and claim their share, however, this 'opportunity' operates as a theft—essentially the parties propose to sell the copyrighted works of absent class members, and then split the proceeds among themselves.

- **It is an unauthorized attempt to revise the rights and remedies of U.S. Copyright law.** "The proposed Settlement Agreement, if approved, would so massively reallocate the existing rights and remedies under copyright law that it would effectively rewrite the existing statutory regime for the benefit of a single player — Google. But Supreme Court precedent is clear: courts may not modify copyright law. Only Congress has 'the constitutional authority and the institutional ability to accommodate fully the varied permutations of competing interests' that must be balanced when amending the Copyright Act.'"

- **It conflicts with international law, specifically The Berne Convention for the Protection of Literary and Artistic Works, an international copyright treaty.** "Not only does the proposed Settlement Agreement attempt to do an end-run around the legislative process, but it also proposes a scheme that Congress could not have adopted because of its clear violation of the United States' international obligations under the Berne Convention for the Protection of Literary and Artistic Works. As Congress has noted, '[a]dherence to [Berne] is in the national interest because it will ensure a strong, credible U.S. presence in the global marketplace...' The Court should not approve what is tantamount to private legislation for the benefit of Google that would violate an international agreement and jeopardize the public's interest in international copyright relations."

- **It gives Google an unlawful and anti-competitive monopoly.** "Finally, because the settlement effectively suspends existing copyright law just for Google, it opens the door for Google to become the dominant player in new markets for online book search engines and book Subscription programs. Accordingly, the settlement should be further rejected because it would violate Section 2 of the Sherman Act, which makes it an offense for any person to "monopolize, or attempt to monopolize, or combine or conspire with any other person or persons, to monopolize any part of the trade or commerce among the several States."

Consumer Watchdog did not bother to raise privacy concerns in the context of the Google Books settlement because we felt the objections just cited were substantial and well enough grounded in law so as to preclude the settlement. Nonetheless other organizations including the Electronic Frontier Foundation, the American Civil Liberties Foundation and the Samuelson Law, Technology & Public Policy Clinic, and a group of privacy authors and publishers have objected on privacy grounds. They said:

> *"Courts, libraries and legislatures have fiercely protected the right to read without fear of being watched or reported upon. The settlement, if approved, may enable Google Book Search to become the world's largest public library, institutional library, book 'purchasing' and ongoing access system combined. It is no understatement to say that this settlement may create the central way that books are accessed in the future, and the only way to access certain books. Because of its potential to greatly expand book access, Google Book Search is extremely exciting.*
>
> *"Yet that future potential will be undermined if this Court allows Google to collect intimate, invasive and previously unavailable information on readers, aggregate that sensitive information with information about them collected by and through other Google products, and by doing so create the real risk of disclosure of that sensitive information to prying governmental entities and private litigants. This chilling effect will hurt authors and publishers, but especially those who write about sensitive or controversial topics. It will also hurt the public interest, as the advance of digitization would come at the cost of reader privacy.*
>
> *"The privacy authors and publishers were not adequately represented in the settlement negotiations. They would not have agreed to a settlement so bereft ofprivacy protections. Without additional protections, the settlement is not fair, reasonable or adequate to the class members or to the public. It should not be approved until sufficient privacy protections are put into place."*

Consumer Watchdog shares those concerns about privacy if a settlement without privacy guarantees is implemented.

So what is to be done? The unique and unfair competitive advantage Google receives under the settlement comes from its attempt to pull an end-run around the appropriate legislative solution to the orphan books problem. This is not an issue for court and certainly one that cannot be settled by solving the problem for one large corporation and no one else.

Congress must resolve the "orphan rights" issue. It could also step in with legislation about what exactly constitutes fair use in the digital age, though that matter could be fairly adjudicated by the courts. Privacy guarantees are another area appropriate for legislative action.

Consumer Watchdog supports digitization and digital libraries in a robust competitive market open to all organizations, both for-profit and non-profit, that offer fundamental privacy guarantees to users. But a single entity cannot be allowed to build a digital library based on a unique monopolistic advantage when its answer to serious questions from responsible critics boils down to: "Trust us. Our motto is 'Don't be evil.'"

Chapter Sources

The following chapters have been previously published:

Chapter 1 – This is an edited, excerpted and augmented edition of a United States Congressional Research Service publication, Report Order Code R40194, dated November 27, 2009.

Chapter 2 – This is an edited, excerpted and augmented edition of a United States Congressional Research Service publication, Report Order Code RL33392, dated February 1, 2010.

Chapter 3 – These remarks were delivered as Statement of Paul Aiken, Executive Director, the Authors Guild, before the Committee on the Judiciary, U.S. House of Representatives, dated September 10, 2009.

Chapter 4 – These remarks were delivered as Statement of Marybeth Peters, The Register of Copyrights, before the Committee on the Judiciary, U.S. House of Representatives, dated September 10, 2009.

Chapter 5 – These remarks were delivered as Statement of David A. Balto, Senior Fellow, Center for American Progress Action Fund, before the Committee on the Judiciary, U.S. House of Representatives, dated September 10, 2009.

Chapter 6 – These remarks were delivered as Statement of David Drummond, Senior Vice President of Corporate Development and Chief Legal Officer, Google Inc., before the Committee on the Judiciary, U.S. House of Representatives, dated September 10, 2009.

Chapter 7 – These remarks were delivered as Statement of Paul Misener, Vice President, Global Public Policy, Amazon.com, before the Committee on the Judiciary, U.S. House of Representatives, dated September 10, 2009.

Chapter 8 – These remarks were delivered as Written Statement of Randall C. Picker, Paul and Theo Leffman Professor of Commercial Law, The University of Chicago Law School and Senior Fellow, The Computation Institute, The University of Chicago and Argonne National Laboratory, before the Committee on the Judiciary, U.S. House of Representatives, dated September 10, 2009.

Chapter 9 – These remarks were delivered as Statement of the National Federation of the the Blind, before the Committee on the Judiciary, U.S. House of Representatives, dated September 10, 2009.

Chapter 10 – These remarks were delivered as Statement of John M. Simpson, Consumer Advocate with Consumer Watchdog, before the Committee on the Judiciary, U.S. House of Representatives, dated September 10, 2009.

INDEX

D

E

F